W0037034

# SpringerBriefs in Cancer Research

For further volumes:
http://www.springer.com/series/10786

Lori Frappier

# EBNA1 and Epstein-Barr Virus Associated Tumours

 Springer

Lori Frappier
University of Toronto
Toronto, ON
Canada

ISBN 978-1-4614-6885-1          ISBN 978-1-4614-6886-8   (eBook)
DOI 10.1007/978-1-4614-6886-8
Springer New York Heidelberg Dordrecht London

Library of Congress Control Number: 2013934383

Printed on acid-free paper

Springer is part of Springer Science+Business Media (www.springer.com)

# Preface

Epstein-Barr Virus (EBV) is a DNA tumour virus that can promote the development of several types B cell lymphomas and carcinomas through its latent mode of infection. EBNA1 was the first EBV protein linked to EBV latent infection and is known to be critical for the persistence of EBV. It is also the only EBV protein expressed in all EBV-associated tumours. EBNA1 is known to play important roles in replicating and partitioning EBV genomes as well as in activating viral gene expression, which would result in indirect contributions to EBV-associated tumours. However in recent years, it has become clear that EBNA1 can also affect cellular proteins and processes in ways that directly implicate EBNA1 as a causative agent in the development and/or persistence of EBV-associated tumours. The multiple potential contributions of EBNA1 to tumourigenesis are reviewed here.

# Contents

# Chapter 1
# Introduction

Epstein-Barr virus (EBV) is a gamma-herpesvirus that infects greater than 90 % of adults worldwide. The virus is spread through saliva and may initially infect the oral epithelium prior to being transmitted to the underlying B-lymphocytes (Rickinson and Kieff 2001). EBV primarily infects resting B lymphocytes and induces their proliferation and B lymphocytes are the main site of persistent EBV infection. Four different types of latent EBV infections have been reported in B cells; three of which occur in proliferating cells and involve expression of viral proteins (latency I, II and III; also called EBNA1-only, default and growth programs), and a fourth that occurs in resting cells in which no viral proteins are expressed (latency 0 or latency program) (Thorley-Lawson and Gross 2004). Epstein-Barr nuclear antigen 1 (EBNA1) is the only EBV protein expressed in all latencies in proliferating cells and the only EBV protein expressed in latency I. It is believed that cells can transition between latency forms, and in particular that latency III results from initial infection and is followed by suppression of viral gene expression resulting in the other latency types. During latent infection, EBV genomes are maintained as circular episomes in the cell nucleus and these undergo replication once every cell cycle to maintain a constant copy number. Latently infected B-cells can switch to a lytic mode of infection that involves turning on a cascade of approximately 80 viral proteins and amplifying the viral genome to produce linear molecules for packaging in virions. This is likely important for transmission of the virus back to the oral epithelium which is a major site of lytic infection and results in viral shedding in the saliva. The interplay between latent and lytic modes of EBV infection results in life-long infection.

While EBV infection is usually asymptomatic, initial EBV infection can result in mononucleosis particularly if initial infection occurs in an adult or adolescent (Rickinson and Kieff 2001). In addition, EBV infection is causally associated with the several types of B-cell lymphomas and carcinomas, including Burkitt's lymphoma, posttransplant lymphoproliferative disease (PTLD), Hodgkin's disease, nasopharyngeal carcinoma (NPC) and gastric carcinomas. In fact, EBV was first identified from Burkitt's lymphoma tumour samples (Epstein et al. 1964).

L. Frappier, *EBNA1 and Epstein-Barr Virus Associated Tumours*,
SpringerBriefs in Cancer Research, DOI: 10.1007/978-1-4614-6886-8_1,
© The Author(s) 2013

EBV-associated tumours have distinct patterns of viral gene expression, resembling latency forms I, II or III. Interestingly, EBNA1 is the only EBV protein that is expressed in all EBV-associated tumours pointing to its importance in EBV cancers. For many years, the importance of EBNA1 was thought to be limited to its requirement for the maintenance of EBV episomes and activation of expression of other EBV genes. However, in the last decade, multiple cellular effects of EBNA1 have been identified that are consistent with a more direct role of EBNA1 in the induction or maintenance of cell transformation. Here, I will review the multiple ways that EBNA1 can potentially contribute to EBV-associated tumours.

## References

Epstein, M. A., Achong, B. G., & Barr, Y. M. (1964). Virus Particles in cultured lymphoblasts from Burkitt's lymphoma. *Lancet, 1*, 702–703.
Rickinson, A. B., & Kieff, E. (2001). Epstein-Barr virus. In D. M. Knipe & P. M. Howley (Eds.), *Fields virology* (pp. 2575–2627). Philadelphia: Lippincott Williams and Wilkins.
Thorley-Lawson, D. A., & Gross, A. (2004). Persistence of the Epstein-Barr virus and the origins of associated lymphomas. *New England Journal of Medicine, 350*, 1328–1337.

# Chapter 2
# Roles of EBNA1 at EBV Episomes

## 2.1 Replication of EBV Episomes

EBV episomes undergo DNA replication once every cell cycle, and therefore resemble the regulated DNA replication typical of eukaryotic cells (Adams 1987). A screen for EBV DNA fragments needed to support the replication of plasmids in cells latently infected with EBV identified a region termed *oriP* (for plasmid origin) that can serve as the origin of DNA replication (Yates et al. 1984). The replication of *oriP* plasmids was shown to require the viral EBNA1 protein (and no additional EBV proteins) (Yates et al. 1985) and to mimic replication of latent EBV episomes in supporting replication that is limited to once per cell cycle (Sternas et al. 1990; Yates and Guan 1991). Subsequent analyses of *oriP* showed that it was comprised two functional elements, termed the dyad symmetry (DS) element and the family of repeats (FR), each of which contained multiple copies of an 18 bp palindromic sequence that was later shown to be bound by EBNA1 (Rawlins et al. 1985; Reisman et al. 1985).

The DS element contains four EBNA1 recognition sites, two of which are located within a 65 bp DS sequence, and three copies of a 9 bp sequence (referred to as nonamers) that is recognised by the host telomeric repeat binding factor 2 (TRF2) (Deng et al. 2002; Niller et al. 1995) (Fig. 2.1). Several studies indicated that the DS is the actual origin of replication within *oriP* and that EBNA1 binding to the DS sequences is essential for its origin function. For example, the DS was shown to be both required and sufficient for plasmid replication in the presence of EBNA1 (Harrison et al. 1994a; Wysokenski and Yates 1989; Yates et al. 2000). In addition, two-dimensional gel analyses of replicating *oriP* plasmids showed that the replication forks initiate from the DS element (Gahn and Schildkraut 1989). While efficient replication from the DS element requires all four EBNA1 binding sites as well as the nonamer repeats, a basal level of DNA replication can be achieved with only two adjacent EBNA1 sites as long as the 3 bp spacing between the sites is retained (Bashaw and Yates 2001; Harrison et al. 1994a, b; Koons et al. 2001; Lindner et al. 2008; Yates et al. 2000). In addition, sequence changes that increase EBNA1 binding affinity result in a corresponding increase in replication

L. Frappier, *EBNA1 and Epstein-Barr Virus Associated Tumours*,
SpringerBriefs in Cancer Research, DOI: 10.1007/978-1-4614-6886-8_2,

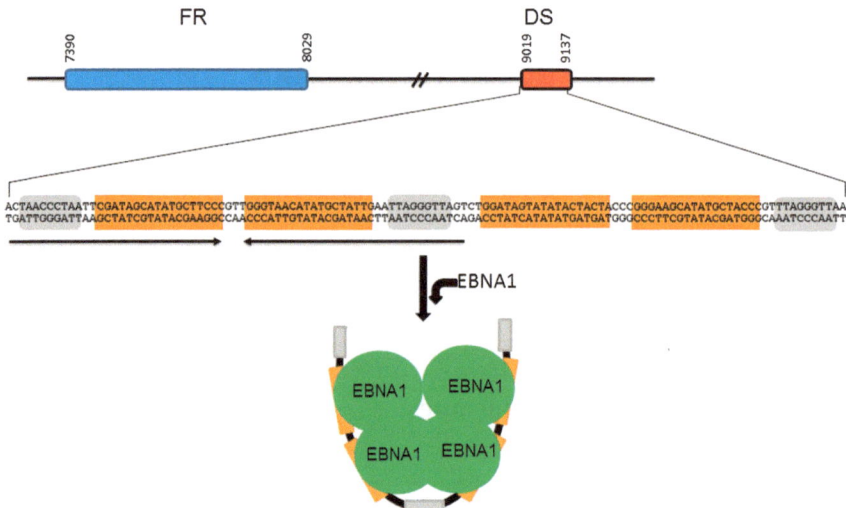

**Fig. 2.1** *Organisation of the oriP DS element.* The location of the *oriP* FR and DS elements in the EBV genome is shown on the *top*, followed by the sequence of the DS element. EBNA1 binding sites in the DS are shown by the *orange boxes* and the nonamer repeats (TRF2 binding sites) are indicated by the *grey boxes*. Converging arrows indicate the 65 bp sequence of dyad symmetry. A model of EBNA1 dimers bound to the DS element is shown on the *bottom*, in which each green circle represents one EBNA1 dimer

efficiency (Lindner et al. 2008). Thus, the EBNA1 interaction with the DS is a major determinant of replication initiation from the DS element.

While EBNA1 binding to the DS is critical for the activation of replication from *oriP*, it is not sufficient for this process. Evidence for this statement comes from the fact that EBNA1 is bound to the *oriP* DS and FR elements throughout the cell cycle, although replication only occurs in S phase (Ritzi et al. 2003). In addition, the EBNA1 DNA binding domain is not sufficient to activate *oriP* replication; rather sequences in the central and N-terminal regions of EBNA1 are also required (Ceccarelli and Frappier 2000; Kim et al. 1997; Kirchmaier and Sugden 1997; Mackey and Sugden 1999; Polvino-Bodnar and Schaffer 1992; Shire et al. 1999; Van Scoy et al. 2000; Wu et al. 2002a; Yates and Camiolo 1988). However, mutational analyses have failed to identify a localised EBNA1 sequence outside of the DNA binding domain that is required for replication, suggesting that multiple N-terminal and central EBNA1 sequences make redundant contributions to replication function. For example, deletion of amino acids 8–67 and 325–376 in combination abrogated EBNA1 replication function (Wu et al. 2002b). Replication activity of EBNA1 can also be increased by deletion or mutation of some EBNA1 sequences; specifically by deletion of residues 61–83 or 395–450 or by point mutation of G81 or G425 (Deng et al. 2005; Holowaty et al. 2003; Wu et al. 2002b). These sequences mediate interactions with the host proteins Brd4 (61–83), USP7 (395–450) and tankyrase (G81 and G425), suggesting that these proteins may negatively regulate replication by EBNA1.

The above observations are consistent with a role for EBNA1 in activating replication by recruiting cellular replication proteins to *oriP*. Moreover, unlike origin binding proteins from several other DNA viruses, EBNA1 lacks DNA helicase activity and is not sufficient to melt the viral origin (Frappier and O'Donnell 1991), further pointing to a need for cellular proteins in initiation events. Indeed, both the cellular origin recognition complex (ORC) and the minichromosome maintenance (MCM) complex were found to be associated with the DS element of *oriP*, implicating them in the initiation and licensing of EBV DNA replication (Chaudhuri et al. 2001; Dhar et al. 2001; Schepers et al. 2001). *OriP* plasmids also failed to replicate stably in a cell line containing a hypomorphic *ORC2* mutation, further supporting the importance of ORC in *oriP* plasmid replication (Dhar et al. 2001). At cellular origins, ORC recruits the MCM complex in conjunction with the action of Cdc6 and Cdt1, and MCM recruitment is inhibited by geminin through its interaction with Cdt1 (Mendez and Stillman 2000; Rialland et al. 2002). Geminin was shown to inhibit replication from *oriP* suggesting that Cdt1 is also involved in MCM recruitment to *oriP* (Dhar et al. 2001). Since the MCM complex is the replicative helicase for cellular DNA replication, it likely fulfils the same role for latent-phase EBV DNA replication.

An important role of EBNA1 in replication from *oriP* is likely in the recruitment of ORC to the DS, as EBNA1 and EBNA1 binding sites in the DS have both been shown to be important for ORC interactions with the DS. In addition, EBNA1 has some ability to interact with ORC (Dhar et al. 2001; Julien et al. 2004; Schepers et al. 2001). EBNA1 was also recently reported to interact with Cdc6 and this interaction increased ORC recruitment to the DS in vitro (Moriyama et al. 2012). Interestingly, while DS-bound EBNA1 recruits ORC, FR-bound EBNA1 has not been found to do so, suggesting that additional feature of the DS DNA contributes to ORC recruitment (Chaudhuri et al. 2001; Dhar et al. 2001; Moriyama et al. 2012; Schepers et al. 2001). Preferential recruitment of ORC to the DS might be explained by the observation that TRF2 bound to the DS nanomers increases ORC recruitment to the DS (Atanasiu et al. 2006; Lindner et al. 2008). However, although nonamer sequences stimulated ORC recruitment, two adjacent EBNA1 binding sites from the DS were shown to be sufficient to recruit ORC in the presence of EBNA1, emphasising the importance of EBNA1 in this process (Atanasiu et al. 2006; Julien et al. 2004). ORC recruitment involves the Gly-Arg-rich regions of EBNA1 (Fig. 2.2), which may interact with ORC through RNA molecules as suggested in work by

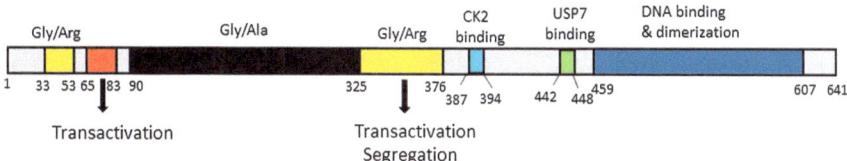

**Fig. 2.2** *Organisation of the EBNA1 protein.* Schematic of the EBNA1 protein showing pertinent regions as discussed in the text and amino acid numbers. The regions outside of the DNA binding domain that are critical for transcriptional activation and segregation are indicated

Norseen et al. (2008, 2009). In particular, G-quadruplex RNA is thought to mediate this interaction as compounds that bind G-quadruplex RNA were found to interfere with the EBNA1-ORC interaction and replication from *oriP* (Norseen et al. 2009). However, another in vitro study suggested that EBNA1 could recruit ORC to the DS in an RNA-independent manner if cdc6 was present (Moriyama et al. 2012).

Another contribution that EBNA1 might make to replication from *oriP* is in altering the chromatin structure of the origin, as EBV genomes are known to be assembled into nucleosomes with a spacing similar to that in cellular chromatin (Shaw et al. 1979). Unlike most sequence–specific DNA binding proteins, EBNA1 was shown to be able to access its recognition sites within the DS when the DS was assembled into a nucleosome or when *oriP* was assembled with physiologically spaced nucleosomes (Avolio-Hunter and Frappier 2003; Avolio-Hunter et al. 2001). EBNA1 destablised nucleosomes on the DS resulting in their displacement. This required EBNA1 interactions with all four recognition sites in the DS and was intrinsic to the EBNA1 DNA binding domain (Avolio-Hunter et al. 2001). Since the four EBNA1 recognitions sites of the DS are also required for efficient DNA replication, nucleosome disruption by EBNA1 may contribute to the efficiency of DNA replication. Indeed, ORC binding to DNA is known to be inhibited by nucleosomes, so nucleosome displacement by EBNA1 might be another way that EBNA1 facilitates the recruitment of ORC and additional replication proteins to the DS. In addition, the assembly of EBNA1 on adjacent sites in the DS is predicted to be accompanied by changes in the DNA structure which might facilitate origin activation (Bochkarev et al. 1996).

EBNA1 may also affect the chromatin structure at *oriP* through the recruitment of template activating factor Iβ (TAF-Iβ also called SET). TAF-Iβ is a nucleosome associated protein that can recruit either histone acetylases or deacetylases to chromatin and has been shown to stimulate the replication and transcription of adenovirus core particles (Matsumoto et al. 1993; Miyamoto et al. 2003; Nagata et al. 1995; Seo et al. 2001; Shikama et al. 2000). EBNA1 binds directly to TAF-Iβ and recruits it to the FR and DS elements of *oriP* (Holowaty et al. 2003; Wang and Frappier 2009). In addition, TAF-Iβ depletion was shown to increase *oriP* plasmid replication while TAF-Iβ overexpression inhibited it, suggesting that TAF-Iβ can negatively regulate replication from *oriP* by affecting the chromatin structure (Wang and Frappier 2009).

The FR element of *oriP* is not required for EBNA1-dependent replication, although it enables *oriP* plasmids to stably persist in proliferating cells by virtue of its mitotic segregation (discussed below). However, the FR can affect replication by inhibiting the movement of replication forks. This inhibition of replication forks requires EBNA1 binding to the multiple tandem sites of the FR and results in a major fork pause site which is detected in both *oriP* plasmids and EBV episomes (Dhar and Schildkraut 1991; Gahn and Schildkraut 1989; Norio and Schildkraut 2001; Norio and Schildkraut 2004). This property of EBNA1 is intrinsic to its DNA binding domain and appears to be due to inhibition of DNA unwinding (Ermakova et al. 1996). It is not clear whether this pausing of replication forks is functionally important or simply a consequence of the complex EBNA1-DNA interaction at this element.

While it is clear that EBNA1 is important to initiate replication from *oriP*, the importance of *oriP* for the replication of EBV episomes is less clear. Questions arose as to the importance of *oriP* for EBV latent genome replication due to replication fork mapping methods that indicated that not all replication forks initiate from *oriP*. Both two-dimensional gel analyses and single molecule analysis of replicated DNA (SMARD) identified two regions of latent EBV episomes that frequently initiate DNA synthesis; *oriP* and a 14 kb region upstream of *oriP* (Little and Schildkraut 1995; Norio and Schildkraut 2001). When oriP was used as the origin, the replication forks initiated from the DS element, consistent with studies on *oriP* plasmids (Norio et al. 2000). In contrast, replication from the 14 kb initiation region was not localised to a single site but initiated from many different sites, resulting in a delocalised initiation pattern reminiscent of the replication initiation zones observed in human genomes (DePamphilis 1993). This region lacks EBNA1 binding sites and therefore is not dependent on EBNA1. While deletion of the DS element from the EBV episome prevented replication from *oriP*, it did not affect replication from the 14 Kb zone, nor did it affect the long-term maintenance of EBV episomes (Norio et al. 2000). This shows that the DS is not absolutely required for EBV episome replication and therefore neither is EBNA1. Subsequent experiments, compared the initiation sites in EBV genomes from Raji and Mutu I Burkitt's lymphoma cells and found that the DS was more frequently used as an origin in Mutu I cells than in Raji cells (Norio and Schildkraut 2004). This variability could be due to the different EBV strains or the different forms of latency involved, since Mutu I cells are latency I (express only EBNA1) and Raji cells are latency III (express all EBNAs and latent membrane proteins). In support of the latter hypothesis, the replication of *oriP* plasmids and EBV genomes in latency I cells was found to be inhibited by expression of the viral latent membrane protein 1 (LMP1) (Shirakata et al. 2001), suggesting that LMP1 expression might be responsible for decreased usage of *oriP* in latency III cells. The results indicate that there is plasticity in origin usage in EBV episomes and that mechanisms exist to replicate EBV episomes in the absence of EBNA1. Consistent with this conclusion, Deutsch et al. (Deutsch et al. 2010) found that mini-EBV genomes lacking the DS could still transform B-lymphocytes and be maintained in these cells, although at a reduced copy number. Therefore, it remains unclear how important the replication function of EBNA1 is for the persistence of EBV episomes.

## 2.2 Segregation of EBV Episomes

Since EBV episomes replicate only once per cell cycle, they require a mechanism to ensure equal segregation or partitioning to the daughter cells during cell division in order for a constant copy number of episomes to be maintained in proliferating cells. Several studies identified EBNA1 and the FR element as critical for stable persistence of *oriP* plasmids in proliferating cells due to effects on mitotic segregation (Krysan et al. 1989; Lee et al. 1999; Lupton and Levine 1985). Interestingly, EBNA1 and the FR can function in segregation independently from the *oriP* DS element as they promote the stable maintenance of a variety of constructs with cellular origins of replication (Kapoor et al. 2001; Krysan et al. 1989; Simpson et al. 1996).

Additional support for the mitotic segregation function of EBNA1 being distinct from its replication function comes from the finding that an EBNA1 mutant lacking the central Gly-Arg-rich region supports the transient replication of *oriP* plasmids at wildtype EBNA1 levels but does not stably maintain them (Shire et al. 1999).

Considerable evidence indicates that EBNA1 mediates segregation by tethering the EBV episomes to the cellular chromosomes in mitosis. This model of segregation initiated with the observations that EBNA1 is tightly associated with the condensed cellular chromosomes in mitosis (Grogan et al. 1983; Harris et al. 1985; Petti et al. 1990). In fact, this was one of the first properties reported for EBNA1. Subsequently, EBV episomes and *oriP*-containing constructs were shown to be associated with mitotic chromosomes and this mitotic chromosome association was found to be dependent on EBNA1 (Delecluse et al. 1993; Harris et al. 1985; Kanda et al. 2001; Kapoor et al. 2005; Simpson et al. 1996). Further evidence that chromosome attachment is important for EBV segregation came from studies examining the EBNA1 sequence requirements for both processes. EBNA1 mutants that are nuclear but defective in mitotic chromosome attachment were found to be impaired in the ability to partition *oriP* plasmids (Hung et al. 2001; Shire et al. 1999; Wu et al. 2000). The EBNA1 segregation function was shown to require both the C-terminal DNA binding domain (for interaction with the EBV episome) and more N-terminal regions for cellular chromosome interactions. The N-terminal regions of EBNA1 can be replaced by chromosome binding sequences from other proteins and the fusion proteins support *oriP* plasmid maintenance, providing strong evidence that chromosome attachment is the contribution of the N-terminal regions to segregation (Hung et al. 2001; Sears et al. 2003).

Studies to map the EBNA1 amino acids important for chromosome attachment began by fusing EBNA1 fragments to GFP and determining which fragments caused GFP to associate with mitotic chromosomes (Marechal et al. 1999). This identified the two Gly-Arg-rich regions of EBNA1 (N-terminal amino acids 33–53 and central amino acids 325–376) as having the ability to associate with mitotic chromosomes (Fig. 2.2). Mutational analyses in the context of the EBNA1 protein then showed that the central Gly-Arg-rich region of EBNA1 was critical for EBNA1 to attach to mitotic chromosomes and to ensure the persistence of *oriP* plasmids (Shire et al. 1999, 2006; Wu et al. 2000, 2002b). However, although a contribution of EBNA1 N-terminal sequences 8–67 was also identified in these studies, deletion of the N-terminal Gly-Arg (33–53) alone did not affect EBNA1's ability to maintain *oriP* plasmids or to associate with mitotic chromosomes (Nayyar et al. 2009; Wu et al. 2002b). Therefore, the central Gly-Arg-rich region of EBNA1 is primarily responsible for mitotic chromosome attachment and EBV partitioning. This region contains a repeated GGRGRGGS sequences that are phosphorylated on the serines and methylated by PRMT1 or PRMT5 on the arginine residues, although the significance of these modifications for EBNA1 functions is not clear (Laine and Frappier 1995; Shire et al. 2006).

Segregation through chromosome attachment is not unique to EBV but is also used by papillomaviruses and by Kaposi sarcoma associated herpesvirus (KSHV), which associate with mitotic chromosomes through their origin binding proteins, E2 and LANA, respectively (Feeney and Parish 2009; You 2010). In each case, the viral proteins associate with mitotic chromosomes through interactions with one or more cellular proteins (Barbera et al. 2006; Krithivas et al. 2002; Parish et al. 2006; You 2010). Of the cellular proteins that have been found to bind EBNA1, one (EBP2 for EBNA1 binding protein 2) has been shown to be important for metaphase chromosome attachment and segregation function of EBNA1 (Kapoor and Frappier 2003; Kapoor et al. 2005; Shire et al. 1999; Wu et al. 2000). The EBNA1 325–376 region important for chromosome attachment was also shown to be critical for EBP2 binding, and mutations within this region were shown to affect EBP2 binding and metaphase chromosome interactions to similar degrees (Nayyar et al. 2009; Shire et al. 1999; Shire et al. 2006; Wu et al. 2000; Wu et al. 2002b). In addition, EBP2 silencing was found to cause EBNA1 and *oriP* plasmids to be released from the metaphase chromosomes (Kapoor et al. 2005). Interestingly, EBP2 was also found to enable EBNA1-mediated plasmid segregation in budding yeast by facilitating EBNA1 attachment to the yeast mitotic chromosomes (Kapoor and Frappier 2003; Kapoor et al. 2001). Studies on the localisation of EBP2 in human cells showed that it is nucleolar in interphase but redistributes to the chromosomes in mitosis (Wu et al. 2000). Detailed studies on the timing of this relocalisation showed that EBP2 is predominantly associated with the chromosomes in the second half of mitosis, whereas EBNA1 is associated with the chromosomes in prophase through telophase (Nayyar et al. 2009). This suggests that the importance of EBP2 is not in mediating the initial attachment of EBNA1 to the chromsomes but rather that EBNA1–EBP2 interaction may be important to maintain EBNA1 on chromosomes. In addition, a recent FRET-based study concluded that the most prevalent EBNA1–EBP2 interactions occured in the nucleoplasm and nucleolus in interphase, raising the possibility that this interphase interaction is somehow important for EBNA–chromosome interactions in mitosis (Jourdan et al. 2012). Interestingly, this possibility is reminiscent of findings for bovine papillomavirus segregation, in which an interphase interaction of the viral E2 protein with the host ChlR1 protein is important for the mitotic chromsome association and segregation functions of E2 (Feeney et al. 2011; Parish et al. 2006).

The initial interaction of EBNA1 with the chromosomes (in prophase or interphase) might involve interactions with other chromosome-associated proteins, direct interactions with the cellular DNA or interactions with chromosome-associated RNA molecules. Support for EBNA1 interactions with cellular DNA or RNA came from observations that the N-terminal and central EBNA1 Gly-Arg-rich regions can interact with DNA or RNA in vitro, perhaps due to their highly basic sequence or their resemblance to AT-hook sequences from other proteins (Norseen et al. 2008, 2009; Sears et al. 2004; Snudden et al., 1994). In particular, EBNA1 has been found to bind G-quadruplex RNA and a compound that binds G-quadruplex RNA was reported to inhibit both the replication function of EBNA1 and its ability to associate with mitotic chromosomes (Norseen et al. 2008, 2009). It is not clear whether the effect of this compound on EBNA1–mitotic chromosome interaction was due to a direct role of G-quadruplex RNA in this process or stems from the inability of the EBV plasmid to replicate in the first place. Note that, in addition to being associated with cellular chromosomes in mitosis, there is evidence that EBNA1 and EBV plasmids are also associated with cellular chromosomes in interphase (Deutsch et al. 2010; Ito et al. 2002; Kanda et al. 2001; Nayyar et al. 2009). In particular, the FR element of *oriP* was found to direct EBV genomes to perichromatic regions with histone modifications typical of active chromatin (Deutsch et al. 2010). Whether or not EBNA1 is mediating this interaction of the FR with active chromatin is not known.

Another question concerning the mechanism of EBNA1-mediated segregation is how equal distribution of the EBV episomes is ensured. Since EBNA1 and EBV episomes are widely distributed over the mitotic chromosomes with no obvious pattern, it was initially suggested that EBNA1 tethered EBV episomes randomly to the chromosomes (Harris et al. 1985). However, random tethering that did not ensure equal distribution on the paired sister chromosomes would not likely result in equal delivery of the EBV episomes to the daughter cells. Indeed, Delecluse et al. (1993) observed that many of the in situ hybridisation signals from EBV episomes were symmetrically located on sister chromatids in Burkitt's lymphoma cells. Additional studies also supported the symmetrical pairing of EBV episomes on daughter chromosomes. Using a recombinant EBV BAC, Kanda et al. (2007) found that EBNA1 and the BAC were often symmetrically distributed on sister chromatids in mitosis. In addition, paired EBNA1 complexes were detected in G2, at which time they were linked through concatenated EBV BACs, a normal product of replication of circular molecules prior to their resolution. Nanbo et al. (2007) also observed a high frequency of paired EBV-based plasmids immediately after DNA replication in G2, that were equally partitioned to daughter cells in M. These studies suggest that the pairing of replicated molecules through catenation may be important for partitioning in mitosis.

## 2.3  Activation of EBV Transcription

Another important function of EBNA1 is in the transcriptional activation of some of the other EBV latency genes, which occurs through the interaction of EBNA1 with the *oriP* FR element. This function was discovered when the presence of the FR element on a plasmid was found to increase the expression of reporter genes in an EBNA1-dependent manner (Lupton and Levine 1985; Reisman and Sugden 1986). This enhancer effect was shown to require 6–7 EBNA1 binding sites from the FR and partial effects were observed with tandem copies of the DS element (Wysokenski and Yates 1989). The EBNA1 and the FR were later shown to activate expression from the viral Cp and LMP promoters suggesting a role for EBNA1 in regulating the expression of the EBNA and LMP EBV latency genes in latent infection (Gahn and Sugden 1995; Sugden and Warren 1989). The importance of this EBNA1 function for EBV infection was confirmed when EBV containing an EBNA1 mutant defective in transcriptional activation (but active for replication and segregation functions) was shown to be severely impaired in the ability to transform cells (Altmann et al. 2006).

The transactivation function of EBNA1 requires its DNA binding domain in order to interact with the FR, but this domain is not sufficient for this function (Polvino-Bodnar and Schaffer 1992). Rather, two additional distinct regions of EBNA1 have been shown to provide the transcriptional activity; the central Gly-Arg-rich region (residues 325–376) and an N-terminal sequence between amino acids 65–83

(Ceccarelli and Frappier 2000; Kennedy and Sugden 2003; Van Scoy et al. 2000; Wang et al. 1997; Wu et al. 2002b; Yates and Camiolo 1988) (Fig. 2.2). The ability of the 325–376 region to activate transcription is inhibited by point mutation of the four serine residues in this sequence, particularly to aspartate, suggesting that these residues participate in transcriptional activation and that phosphorylation of these serines might negatively regulate this activity (Shire et al. 2006). As mentioned above, the 325–376 region also plays a critical role in the segregation function of EBNA1, and it is not known whether the same or different sets of interactors with this region mediate these two functions. No detailed mutational analyses have been conducted on the 65–83 EBNA1 region but the discovery that deletion of this region disrupted the transcriptional function of EBNA1 without affecting the replication or segregation functions clearly showed that transcriptional activation is a distinct function of EBNA1 (Wu et al. 2002a). In addition, deletion of EBNA1 residues 65–89 in the context of an infectious EBV was shown to be defective in activating expression of the EBNA genes from the Cp promoter and hence impaired in cell transformation, although still capable of supporting stable plasmid replication (Altmann et al. 2006).

The finding that both of the EBNA1 transcriptional activation sequences are required for efficient transactivation, suggests that they contribute in distinct ways to transcription, most likely by mediating different cellular protein interactions. The 65–83 region of EBNA1 has been found to interact with tankyrase (Deng et al. 2005) and with Nm23-H1 (Murakami et al. 2005 see below); however neither of these proteins have been linked to the transcription function of EBNA1. This region was also found to mediate an interaction with Brd4 (Lin et al. 2008), a cellular bromodomain protein that interacts with chromatin to regulate transcription (Wu and Chiang 2007). Links between Brd4 and EBNA1-mediated transcription were also revealed from the findings that Brd4 depletion inhibited EBNA1-mediated transcriptional activation and that Brd4 localised with EBNA1 at the FR element (but not at the DS) (Lin et al., 2008). Therefore, Brd4 is currently the best candidate for mediating the transcriptional effect of the EBNA1 65–83 transactivation region. Interestingly, the papillomavirus E2 proteins, which carryout many of the same functions as EBNA1, also interact with Brd4 and use this interaction for transcriptional activation (Ilves et al. 2006; McPhillips et al. 2006; Schweiger et al. 2006). Therefore, EBNA1 and E2 may use common mechanisms to activate transcription.

The EBNA1 325–376 region mediates a few different host protein interactions that may impact transcription. Several groups have reported an interaction between EBNA1 and P32/TAP (also called gC1q) and this interaction is largely mediated by the 325–376 region (Holowaty et al. 2003; Ito et al. 2000; Shire et al. 1999; Van Scoy et al. 2000; Wang et al. 1997). P32/TAP is mainly mitochondrial but due to its highly acid nature, interacts with a wide variety of arg-rich proteins (Jiang et al. 1999). However, a role for this protein in transcriptional activation by EBNA1 was suggested, because it was detected at *oriP* by chromatin immunoprecipitation and the C-terminal fragment of P32/TAP was found to activate a reporter gene when fused to the GAL4 DNA binding domain (Van Scoy et al. 2000; Wang et al. 1997). Two related nucleosome assembly proteins with clear ties to transcriptional regulation, NAP1 and TAF-Iβ (also called SET), also interact with the EBNA1 325–376 sequence (Holowaty et al. 2003; Park and Luger 2006; Wang and Frappier 2009). Both NAP1 and TAF-Iβ have been shown to activate transcription of adenovirus core particles (Haruki et al. 2006; Kawase et al. 1996; Matsumoto et al. 1995) and NAP1 has been shown to be used by papillomavirus E2 to activate transcription through recruitment of the p300 histone acetyltransferase (Rehtanz et al. 2004; Shikama et al. 2000). In addition, both proteins have been implicated in cellular gene expression through effects on histone acetylation (Kutney et al. 2004; Miyamoto et al. 2003; Park and Luger 2006; Seo et al. 2001; Shikama et al. 2000). Support for a role of NAP1 and TAF-Iβ in EBNA1-mediated transcriptional activation comes from the finding that both proteins are recruited to the FR element by EBNA1 and that silencing of either NAP1 or TAF-Iβ decreases EBNA1 transactivation activity (Wang and Frappier 2009).

Finally, the state of histone H2B ubiquitylation may also influence transcriptional activation by EBNA1. This conclusion stems from the finding that EBNA1 recruits a H2B deubiquitylation complex, consisting of USP7 and GMP synthetase, to the FR through a direct interaction with USP7 (Sarkari et al. 2009). USP7 silencing was found to increase the level of monoubiquitylated H2B at the FR and to decrease transcriptional activation by EBNA1, suggesting that monoubiquitylation of H2B inhibits EBNA1-mediated transactivation. In keeping with this result, an EBNA1 mutant defective in USP7 binding was found to have decreased transcriptional activation activity (Holowaty et al. 2003).

# References

Adams, A. (1987). Replication of latent Epstein-Barr virus genomes. *Journal of Virology, 61,* 1743–1746.

Altmann, M., Pich, D., Ruiss, R., Wang, J., Sugden, B., & Hammerschmidt, W. (2006). Transcriptional activation by EBV nuclear antigen 1 is essential for the expression of EBV's transforming genes. *Proceedings of National Academy of Sciences United States of America, 103,* 14188–14193.

Atanasiu, C., Deng, Z., Wiedmer, A., Norseen, J., & Lieberman, P. M. (2006). ORC binding to TRF2 stimulates OriP replication. *EMBO Reports, 7,* 716–721.

Avolio-Hunter, T. M., & Frappier, L. (2003). EBNA1 efficiently assembles on chromatin containing the Epstein-Barr virus latent origin of replication. *Virology, 315,* 398–408.

Avolio-Hunter, T. M., Lewis, P. N., & Frappier, L. (2001). Epstein-Barr nuclear antigen 1 binds and destbilizes nucleosomes at the viral origin of latent DNA replication. *Nucleic Acids Research, 29,* 3520–3528.

Barbera, A. J., Chodaparambil, J. V., Kelley-Clarke, B., Joukov, V., Walter, J. C., Luger, K., et al. (2006). The nucleosomal surface as a docking station for Kaposi's sarcoma herpesvirus LANA. *Science, 311,* 856–861.

Bashaw, J. M., & Yates, J. L. (2001). Replication from oriP of Epstein-Barr virus requires exact spacing of two bound dimers of EBNA1 which bend DNA. *Journal of Virology, 75,* 10603–10611.

Bochkarev, A., Barwell, J., Pfuetzner, R., Bochkareva, E., Frappier, L., & Edwards, A. M. (1996). Crystal structure of the DNA-binding domain of the Epstein-Barr virus origin binding protein, EBNA1, bound to DNA. *Cell, 84,* 791–800.

Ceccarelli, D. F. J., & Frappier, L. (2000). Functional analyses of the EBNA1 origin DNA binding protein of Epstein-Barr virus. *Journal of Virology, 74,* 4939–4948.

Chaudhuri, B., Xu, H., Todorov, I., Dutta, A., & Yates, J. L. (2001). Human DNA replication initiation factors, ORC and MCM, associate with *oriP* of Epstein-Barr virus. *Proceedings of the National Academy of Sciences of the United States of America, 98*, 10085–10089.

Delecluse, H.-J., Bartnizke, S., Hammerschmidt, W., Bullerdiek, J., & Bornkamm, G. W. (1993). Episomal and integrated copies of Epstein-Barr virus coexist in Burkitt's lymphoma cell lines. *Journal of Virology, 67*, 1292–1299.

Deng, Z., Atanasiu, C., Zhao, K., Marmorstein, R., Sbodio, J. I., Chi, N. W., et al. (2005). Inhibition of Epstein-Barr virus OriP function by tankyrase, a telomere-associated poly-ADP ribose polymerase that binds and modifies EBNA1. *Journal of Virology, 79*, 4640–4650.

Deng, Z., Lezina, L., Chen, C.-J., Shtivelband, S., So, W., & Lieberman, P. M. (2002). Telomeric proteins regulate episomal maintenance of Epstein-Barr virus origin of plasmid replication. *Molecular Cell, 9*, 493–503.

DePamphilis, M. L. (1993). Eukaryotic DNA replication: Anatomy of an origin. *Annual Review of Biochemistry, 62*, 29–63.

Deutsch, M. J., Ott, E., Papior, P., & Schepers, A. (2010). The latent origin of replication of Epstein-Barr virus directs viral genomes to active regions of the nucleus. *Journal of Virology, 84*, 2533–2546.

Dhar, V., & Schildkraut, C. L. (1991). Role of EBNA-1 in arresting replication forks at the Epstein-Barr virus *oriP* family of tandem repeats. *Molecular and Cellular Biology, 11*, 6268–6278.

Dhar, S. K., Yoshida, K., Machida, Y., Khaira, P., Chaudhuri, B., Wohlschlegel, J. A., et al. (2001). Replication from oriP of Epstein-Barr virus requires human ORC and is inhibited by geminin. *Cell, 106*, 287–296.

Ermakova, O., Frappier, L., & Schildkraut, C. L. (1996). Role ot the EBNA-1 protein in pausing of replication forks in the Epstein-Barr virus genome. *Journal of Biological Chemistry, 271*, 33009–33017.

Feeney, K. M., & Parish, J. L. (2009). Targeting mitotic chromosomes: A conserved mechanism to ensure viral genome persistence. *Proceedings Biological Sciences, 276*, 1535–1544.

Feeney, K. M., Saade, A., Okrasa, K., & Parish, J. L. (2011). In vivo analysis of the cell cycle dependent association of the bovine papillomavirus E2 protein and ChlR1. *Virology, 414*, 1–9.

Frappier, L., & O'Donnell, M. (1991). Overproduction, purification and characterization of EBNA1, the origin binding protein of Epstein-Barr virus. *Journal of Biological Chemistry, 266*, 7819–7826.

Gahn, T. A., & Schildkraut, C. L. (1989). The Epstein-Barr virus origin of plasmid replication, *oriP*, contains both the initiation and termination sites of DNA replication. *Cell, 58*, 527–535.

Gahn, T., & Sugden, B. (1995). An EBNA1 dependent enhancer acts from a distance of 10 kilobase pairs to increase expression of the Epstien-Barr virus LMP gene. *Journal of Virology, 69*, 2633–2636.

Grogan, E. A., Summers, W. P., Dowling, S., Shedd, D., Gradoville, L., & Miller, G. (1983). Two Epstein-Barr viral nuclear neoantigens distinguished by gene transfer, serology and chromosome binding. *Proceedings of the National Academy of Sciences of the United States of America, 80*, 7650–7653.

Harris, A., Young, B. D., & Griffin, B. E. (1985). Random association of Epstein-Barr virus genomes with host cell metaphase chromosomes in Burkitt's lymphoma-derived cell lines. *Journal of Virology, 56*, 328–332.

Harrison, S., Fisenne, K., & Hearing, J. (1994a). Sequence requirements of the Epstein-Barr Virus latent origin of DNA replication. *Journal of Virology, 68*, 1913–1925.

Harrison, S., Fisenne, K., & Hearing, J. (1994b). Sequence requirements of the Epstein-Barr virus latent origin of DNA replication. *Journal of Virology, 68*, 1913–1925.

Haruki, H., Okuwaki, M., Miyagishi, M., Taira, K., & Nagata, K. (2006). Involvement of template-activating factor I/SET in transcription of adenovirus early genes as a positive-acting factor. *Journal of Virology, 80*, 794–801.

Holowaty, M. N., Zeghouf, M., Wu, H., Tellam, J., Athanasoupoulos, V., Greenblatt, J., et al. (2003). Protein profiling with Epstein-Barr nuclear antigen-1 reveals an interaction with the herpesvirus-associated ubiquitin-specific protease HAUSP/USP7. *Journal of Biological Chemistry, 278*, 29987–29994.

Hung, S. C., Kang, M.-S., & Kieff, E. (2001). Maintenance of Epstein-Barr virus (EBV) oriP-based episomes requires EBV-encoded nuclear antigen-1 chromosome-binding domains, which can be replaced by high-mobility group-I or histone H1. *Proceedings of the National Academy of Sciences of the United States of America, 98*, 1865–1870.

Ilves, I., Maemets, K., Silla, T., Janikson, K., & Ustav, M. (2006). Brd4 is involved in multiple processes of the bovine papillomavirus type 1 life cycle. *Journal of Virology, 80*, 3660–3665.

Ito, S., Gotoh, E., Ozawa, S., & Yanagi, K. (2002). Epstein-Barr virus nuclear antigen-1 is highly colocalized with interphase chromatin and its newly replicated regions in particular. *Journal of General Virology, 83*, 2377–2383.

Ito, S., Ikeda, M., Kato, N., Matsumoto, A., Ishikawa, Y., Kumakubo, S., et al. (2000). Epstein-Barr virus nuclear antigen-1 binds to nuclear transporter karyopherin α1/NPI-1 in addition to karyopherin α2/Rch1. *Virology, 266*, 110–119.

Jiang, J., Zhang, Y., Krainer, A. R., & Xu, R. M. (1999). Crystal structure of human p32, a doughnut-shaped acidic mitochondrial matrix protein. *Proceedings of the National Academy of Sciences of the United States of America, 96*, 3572–3577.

Jourdan, N., Jobart-Malfait, A., Dos Reis, G., Quignon, F., Piolot, T., Klein, C., et al. (2012). Live-cell imaging reveals multiple interactions between Epstein-Barr virus nuclear antigen 1 and cellular chromatin during interphase and mitosis. *Journal of Virology, 86*, 5314–5329.

Julien, M. D., Polonskaya, Z., & Hearing, J. (2004). Protein and sequence requirements for the recruitment of the human origin recognition complex to the latent cycle origin of DNA replication of Epstein-Barr virus oriP. *Virology, 326*, 317–328.

Kanda, T., Kamiya, M., Maruo, S., Iwakiri, D., & Takada, K. (2007). Symmetrical localization of extrachromosomally replicating viral genomes on sister chromatids. *Journal of Cell Science, 120*, 1529–1539.

Kanda, T., Otter, M., & Wahl, G. M. (2001). Coupling of mitotic chromosome tethering and replication competence in Epstein-Barr virus-based plasmids. *Molecular and Cellular Biology, 21*, 3576–3588.

Kapoor, P., & Frappier, L. (2003). EBNA1 partitions Epstein-Barr virus plasmids in yeast by attaching to human EBNA1-binding protein 2 on mitotic chromosomes. *Journal of Virology, 77*, 6946–6956.

Kapoor, P., Lavoie, B. D., & Frappier, L. (2005). EBP2 plays a key role in Epstein-Barr virus mitotic segregation and is regulated by aurora family kinases. *Molecular and Cellular Biology, 25*, 4934–4945.

Kapoor, P., Shire, K., & Frappier, L. (2001). Reconstitution of Epstein-Barr virus-based plasmid partitioning in budding yeast. *EMBO Journal, 20*, 222–230.

Kawase, H., Okuwaki, M., Miyaji, M., Ohba, R., Handa, H., Ishimi, Y., et al. (1996). NAP-1 is a functional homologue of TAF-I that is required for replication and transcription of the adenovirus genome in a chromatin-like structure. *Genes to Cells, 1*, 1045–1056.

Kennedy, G., & Sugden, B. (2003). EBNA-1, a bifunctional transcriptional activator. *Molecular and Cellular Biology, 23*, 6901–6908.

Kim, A. L., Maher, M., Hayman, J. B., Ozer, J., Zerby, D., Yates, J. L., et al. (1997). An imperfect correlation between DNA replication activity of Epstein-Barr virus nuclear antigen 1 (EBNA1) and binding to the nuclear import receptor, Rch1/importin α. *Virology, 239*, 340–351.

Kirchmaier, A. L., & Sugden, B. (1997). Dominant-negative inhibitors of EBNA1 of Epstein-Barr virus. *Journal of Virology, 71*, 1766–1775.

Koons, M. D., Van Scoy, S., & Hearing, J. (2001). The replicator of the Epstein-Barr virus latent cycle origin of DNA replication, *oriP*, is composed of multiple functional elements. *Journal of Virology, 75*, 10582–10592.

Krithivas, A., Fujimuro, M., Weidner, M., Young, D. B., & Hayward, S. D. (2002). Protein interactions targeting the latency-associated nuclear antigen of Kaposi's sarcoma-associated herpesvirus to cell chromosomes. *Journal of Virology, 76*, 11596–11604.

Krysan, P. J., Haase, S. B., & Calos, M. P. (1989). Isolation of human sequences that replicate autonomously in human cells. *Molecular and Cellular Biology, 9*, 1026–1033.

Kutney, S. N., Hong, R., Macfarlan, T., & Chakravarti, D. (2004). A signaling role of histone-binding proteins and INHAT subunits pp 32 and Set/TAF-Ibeta in integrating chromatin hypoacetylation and transcriptional repression. *Journal of Biological Chemistry, 279*, 30850–30855.

Laine, A., & Frappier, L. (1995). Identification of Epstein-Barr nuclear antigen 1 protein domains that direct interactions at a distance between DNA-bound proteins. *Journal of Biological Chemistry, 270*, 30914–30918.

Lee, M. A., Diamond, M. E., & Yates, J. L. (1999). Genetic evidence that EBNA-1 is needed for efficient, stable latent infection by Epstein-Barr virus. *Journal of Virology, 73*, 2974–2982.

Lin, A., Wang, S., Nguyen, T., Shire, K., & Frappier, L. (2008). The EBNA1 protein of Epstein-Barr virus functionally interacts with Brd4. *Journal of Virology, 82*, 12009–12019.

Lindner, S. E., Zeller, K., Schepers, A., & Sugden, B. (2008). The affinity of EBNA1 for its origin of DNA synthesis is a determinant of the origin's replicative efficiency. *Journal of Virology, 82*, 5693–5702.

Little, R. D., & Schildkraut, C. L. (1995). Initiation of latent DNA replicatoon in the Epstein-Barr virus genome can occur at sites other than the genetically defined origin. *Molecular and Cellular Biology, 15*, 2893–2903.

Lupton, S., & Levine, A. J. (1985). Mapping of genetic elements of Epstein-Barr virus that facilitate extrachromosomal persistence of Epstein-Barr virus-derived plasmids in human cells. *Molecular and Cellular Biology, 5*, 2533–2542.

Mackey, D., & Sugden, B. (1999). The linking regions of EBNA1 are essential for its support of replication and transcription. *Molecular and Cellular Biology, 19*, 3349–3359.

Marechal, V., Dehee, A., Chikhi-Brachet, R., Piolot, T., Coppey-Moisan, M., & Nicolas, J. (1999). Mapping EBNA1 domains involved in binding to metaphase chromosomes. *Journal of Virology, 73*, 4385–4392.

Matsumoto, K., Nagata, K., Ui, M., & Hanaoka, F. (1993). Template activating factor I, a novel host factor required to stimulate the adenovirus core DNA replication. *Journal of Biological Chemistry, 268*, 10582–10587.

Matsumoto, K., Okuwaki, M., Kawase, H., Handa, H., Hanaoka, F., & Nagata, K. (1995). Stimulation of DNA transcription by the replication factor from the adenovirus genome in a chromatin-like structure. *Journal of Biological Chemistry, 270*, 9645–9650.

McPhillips, M. G., Oliveira, J. G., Spindler, J. E., Mitra, R., & McBride, A. A. (2006). Brd4 is required for E2-mediated transcriptional activation but not genome partitioning of all papillomaviruses. *Journal of Virology, 80*, 9530–9543.

Mendez, J., & Stillman, B. (2000). Chromatin association of human origin recognition coplex, cdc6, and minichromosome maintenance proteins during the cell cycle: assembly of prereplication complexes in late mitosis. *Molecular and Cellular Biology, 20*, 8602–8612.

Miyamoto, S., Suzuki, T., Muto, S., Aizawa, K., Kimura, A., Mizuno, Y., et al. (2003). Positive and negative regulation of the cardiovascular transcription factor KLF5 by p300 and the oncogenic regulator SET through interaction and acetylation on the DNA-binding domain. *Molecular and Cellular Biology, 23*, 8528–8541.

Moriyama, K., Yoshizawa-Sugata, N., Obuse, C., Tsurimoto, T., & Masai, H. (2012). Epstein-Barr nuclear antigen 1 (EBNA1)-dependent recruitment of origin recognition complex (Orc) on oriP of Epstein-Barr virus with purified proteins: Stimulation by Cdc6 through its direct interaction with EBNA1. *Journal of Biological Chemistry, 287*, 23977–23994.

Murakami, M., Lan, K., Subramanian, C., & Robertson, E. S. (2005). Epstein-Barr virus nuclear antigen 1 interacts with Nm23-H1 in lymphoblastoid cell lines and inhibits its ability to suppress cell migration. *Journal of Virology, 79*, 1559–1568.

Nagata, K., Kawase, H., Handa, H., Yano, K., Yamasaki, M., Ishimi, Y., et al. (1995). Replication factor encoded by a putative oncogene, set, associated with myeloid leukemogenesis. *Proceedings of the National Academy of Sciences of the United States of America, 92*, 4279–4283.

Nanbo, A., Sugden, A., & Sugden, B. (2007). The coupling of synthesis and partitioning of EBV's plasmid replicon is revealed in live cells. *EMBO Journal, 26*, 4252–4262.

Nayyar, V. K., Shire, K., & Frappier, L. (2009). Mitotic chromosome interactions of Epstein-Barr nuclear antigen (EBNA1) and human EBNA1-binding protein 2 (EBP2). *Journal of Cell Science, 122*, 4341–4350.

Niller, H. H., Glaser, G., Knuchel, R., & Wolf, H. (1995). Nucleoprotein complexes and DNA 5'-ends at oriP of Epstein-Barr virus. *Journal of Biological Chemistry, 270*, 12864–12868.

Norio, P., & Schildkraut, C. L. (2001). Visualization of DNA replication on individual Epstein-Barr virus episomes. *Science, 294*, 2361–2364.

Norio, P., & Schildkraut, C. L. (2004). Plasticity of DNA replication initiation in Epstein-Barr virus episomes. *PLoS Biology, 2*, e152.

Norio, P., Schildkraut, C. L., & Yates, J. L. (2000). Initiation of DNA replication within *oriP* is dispensable for stable replication of the latent Epstein-Barr virus chromosome after infection of established cell lines. *Journal of Virology, 74*, 8563–8574.

Norseen, J., Johnson, F. B., & Lieberman, P. M. (2009). Role for G-quadruplex RNA binding by Epstein-Barr virus nuclear antigen 1 in DNA replication and metaphase chromosome attachment. *Journal of Virology, 83*, 10336–10346.

Norseen, J., Thomae, A., Sridharan, V., Aiyar, A., Schepers, A., & Lieberman, P. M. (2008). RNA-dependent recruitment of the origin recognition complex. *EMBO Journal, 27*, 3024–3035.

Parish, J. L., Bean, A. M., Park, R. B., & Androphy, E. J. (2006). ChlR1 is required for loading papillomavirus E2 onto mitotic chromosomes and viral genome maintenance. *Molecular Cell, 24*, 867–876.

Park, Y. J., & Luger, K. (2006). Structure and function of nucleosome assembly proteins. *Biochemistry and Cell Biology, 84*, 549–558.

Petti, L., Sample, C., & Kieff, E. (1990). Subnuclear localization and phosphorylation or Epstein-Barr virus latent infection nuclear proteins. *Virology, 176*, 563–574.

Polvino-Bodnar, M., & Schaffer, P. A. (1992). DNA binding activity is required for EBNA1-dependent transcriptional activation and DNA replication. *Virology, 187*, 591–603.

Rawlins, D. R., Milman, G., Hayward, S. D., & Hayward, G. S. (1985). Sequence-specific DNA binding of the Epstein-Barr virus nuclear antigen (EBNA1) to clustered sites in the plasmid maintenance region. *Cell, 42*, 859–868.

Rehtanz, M., Schmidt, H. M., Warthorst, U., & Steger, G. (2004). Direct interaction between nucleosome assembly protein 1 and the papillomavirus E2 proteins involved in activation of transcription. *Molecular and Cellular Biology, 24*, 2153–2168.

Reisman, D., & Sugden, B. (1986). *trans* Activation of an Epstein-Barr viral transcriptional enhancer by the Epstein-Barr viral nuclear antigen 1. *Molecular and Cellular Biology, 6*, 3838–3846.

Reisman, D., Yates, J., & Sugden, B. (1985). A putative origin of replication of plasmids derived from Epstein-Barr virus is composed of two cis-acting components. *Molecular and Cellular Biology, 5*, 1822–1832.

Rialland, M., Sola, F., & Santocanale, C. (2002). Essential role of human CDT1 in DNA replication and chromatin licensing. *Journal of Cell Science, 115*, 1435–1440.

Ritzi, M., Tillack, K., Gerhardt, J., Ott, E., Humme, S., Kremmer, E., et al. (2003). Complex protein-DNA dynamics at the latent origin of DNA replication of Epstein-Barr virus. *Journal of Cell Science, 116*, 3971–3984.

Sarkari, F., Sanchez-Alcaraz, T., Wang, S., Holowaty, M. N., Sheng, Y., & Frappier, L. (2009). EBNA1-mediated recruitment of a histone H2B deubiquitylating complex to the Epstein-Barr virus latent origin of DNA replication. *PLoS Pathogens, 5*, e1000624.

Schepers, A., Ritzi, M., Bousset, K., Kremmer, E., Yates, J. L., Harwood, J., et al. (2001). Human origin recognition complex binds to the region of the latent origin of DNA replication of Epstein-Barr virus. *EMBO Journal, 20*, 4588–4602.

Schweiger, M. R., You, J., & Howley, P. M. (2006). Bromodomain protein 4 mediates the papillomavirus E2 transcriptional activation function. *Journal of Virology, 80*, 4276–4285.

Sears, J., Kolman, J., Wahl, G. M., & Aiyar, A. (2003). Metaphase chromosome tethering is necessary for the DNA synthesis and maintenance of oriP plasmids but is insufficient for transcription activation by Epstein-Barr nuclear antigen 1. *Journal of Virology, 77*, 11767–11780.

Sears, J., Ujihara, M., Wong, S., Ott, C., Middeldorp, J., & Aiyar, A. (2004). The amino terminus of Epstein-Barr Virus (EBV) nuclear antigen 1 contains AT hooks that facilitate the replication and partitioning of latent EBV genomes by tethering them to cellular chromosomes. *Journal of Virology, 78*, 11487–11505.

Seo, S.-B., McNamara, P., Heo, S., Turner, A., Lane, W. S., & Chakravarti, D. (2001). Regulation of histone acetylation and transcription by INHAT, a human cellular complex containing the Set oncoprotein. *Cell, 104*, 119–130.

Shaw, J., Levinger, L., & Carter, C. (1979). Nucleosomal structure of Epstein-Barr virus DNA in transformed cell lines. *Journal of Virology, 29*, 657–665.

Shikama, N., Chan, H. M., Krstic-Demonacos, M., Smith, L., Lee, C. W., Cairns, W., et al. (2000). Functional interaction between nucleosome assembly proteins and p300/CREB-binding protein family coactivators. *Molecular and Cellular Biology, 20*, 8933–8943.

Shirakata, M., Imadome, K.-I., Okazaki, K., & Hirai, K. (2001). Activation of TRAF5 and TRAF6 signal cascades negatively regulates the latent replication origin of Epstein-Barr virus through p38 mitogen-activated protein kinase. *Journal of Virology, 75*, 5059–5068.

Shire, K., Ceccarelli, D. F. J., Avolio-Hunter, T. M., & Frappier, L. (1999). EBP2, a human protein that interacts with sequences of the Epstein-Barr nuclear antigen 1 important for plasmid maintenance. *Journal of Virology, 73*, 2587–2595.

Shire, K., Kapoor, P., Jiang, K., Hing, M. N., Sivachandran, N., Nguyen, T., et al. (2006). Regulation of the EBNA1 Epstein-Barr virus protein by serine phosphorylation and arginine methylation. *Journal of Virology, 80*, 5261–5272.

Simpson, K., McGuigan, A., & Huxley, C. (1996). Stable episomal maintenance of yeast artificial chromosomes in human cells. *Molecular and Cellular Biology, 16*, 5117–5126.

Snudden, D.K., Hearing, J., Smith, P.R., Grasser, F.A., & Griffin, B.E. (1994). EBNA1, the major nuclear antigen of Epstein-Barr virus, resenbles 'RGG' RNA binding proteins. *EMBO Journal, 13*, 4840-4848-4847.

Sternas, L., Middleton, T., & Sugden, B. (1990). The average number of molecules of Epstein-Barr nuclear antigen 1 per cell does not correlate with the average number of Epstein-Barr virus (EBV) DNA molecules per cell among different clones of EBV-immortalized cells. *Journal of Virology, 64*, 2407–2410.

Sugden, B., & Warren, N. (1989). A promoter of Epstein-Barr virus that can function during latent infection can be transactivated by EBNA-1, a viral protein required for viral DNA replication during latent infection. *Journal of Virology, 63*, 2644–2649.

Van Scoy, S., Watakabe, I., Krainer, A. R., & Hearing, J. (2000). Human p32: A coactivator for Epstein-Barr virus nuclear antigen-1-mediated transcriptional activation and possible role in viral latent cycle DNA replication. *Virology, 275*, 145–157.

Wang, Y., Finan, J. E., Middeldorp, J. M., & Hayward, S. D. (1997). P32/TAP, a cellular protein that interacts with EBNA-1 of Epstein-Barr virus. *Virology, 236*, 18–29.

Wang, S., & Frappier, L. (2009). Nucleosome assembly proteins bind to Epstein-Barr virus nuclear antigen 1 and affect its functions in DNA replication and transcriptional activation. *Journal of Virology, 83*, 11704–11714.

Wu, H., Ceccarelli, D. F. J., & Frappier, L. (2000). The DNA segregation mechanism of the Epstein-Barr virus EBNA1 protein. *EMBO Reports, 1*, 140–144.

Wu, S. Y., & Chiang, C. M. (2007). The double bromodomain-containing chromatin adaptor Brd4 and transcriptional regulation. *Journal of Biological Chemistry, 282*, 13141–13145.

Wu, H., Kapoor, P., & Frappier, L. (2002a). Separation of the DNA replication, segregation and transcriptional activation functions of Epstein-Barr nuclear antigen 1. *Journal of Virology, 76*, 2480–2490.

Wu, H., Kapoor, P., & Frappier, L. (2002b). Separation of the DNA replication, segregation, and transcriptional activation functions of Epstein-Barr nuclear antigen 1. *Journal of Virology, 76*, 2480–2490.

Wysokenski, D. A., & Yates, J. L. (1989). Multiple EBNA1-binding sites are required to form an EBNA1-dependent enhancer and to activate a minimal replicative origin within *oriP* of Epstein-Barr virus. *Journal of Virology, 63*, 2657–2666.

Yates, J. L., & Camiolo, S. M. (1988). Dissection of DNA replication and enhancer activation functions of Epstein-Barr virus nuclear antigen 1. *Cancer Cells, 6*, 197–205.

Yates, J. L., Camiolo, S. M., & Bashaw, J. M. (2000). The minimal replicator of Epstein-Barr virus oriP. *Journal of Virology, 74*, 4512–4522.

Yates, J. L., & Guan, N. (1991). Epstein-Barr virus-derived plasmids replicate only once per cell cycle and are not amplified after entry into cells. *Journal of Virology, 65,* 483–488.

Yates, J. L., Warren, N., Reisman, D., & Sugden, B. (1984). A cis-acting element from the Epstein-Barr viral genome that permits stable replication of recombinant plasmids in latently infected cells. *Proceedings of the National Academy of Sciences of the United States of America, 81,* 3806–3810.

Yates, J. L., Warren, N., & Sugden, B. (1985). Stable replication of plasmids derived from Epstein-Barr virus in various mammalian cells. *Nature, 313,* 812–815.

You, J. (2010). Papillomavirus interaction with cellular chromatin. *Biochimica et Biophysica Acta, 1799,* 192–199.

# Chapter 3
# EBNA1-DNA Interactions

## 3.1 Mechanism of EBNA1 Interaction with EBV DNA

Soon after EBNA1 was found to be required to replicate and maintain *oriP* plasmids, EBNA1 was shown to bind to specific sequences in the *oriP* FR and DS elements through its C-terminal region (Jones et al. 1989; Rawlins et al. 1985). A screen for EBV genomic fragments bound by EBNA1 not only identified the FR and DS elements but also a fragment referred to as BamHI-Q (Jones et al. 1989), which was later shown to contain a promoter (Qp) used for the expression of EBNA1 in the absence of other EBNAs (Jones et al. 1989; Nonkwelo et al. 1996; Sample et al. 1992). EBNA1 was shown to have the highest affinity for the FR region, followed by the DS then the BamHI-Q region (Jones et al. 1989). All of the EBNA1-bound EBV fragments contained multiple copies of an 18 bp palindromic sequence that was protected by EBNA1 (Ambinder et al. 1990; Rawlins et al. 1985). The multiple copies of this sequence in the EBV genome are not identical but contain some variations that account for the different affinities of EBNA1 for the FR, DS and BamHI-Q regions and for individual sites within these regions (Ambinder et al. 1990; Summers et al. 1996). The BamHI-Q region contains two EBNA1 recognition sites that decrease EBNA1 expression when bound by EBNA1, enabling EBNA1 to autoregulate its own expression (Ambinder et al. 1990; Jones et al. 1989; Yoshioka et al. 2008).

The EBNA1 domain responsible for DNA binding lies between amino acids 459 and 607, and the same domain mediates the dimerisation of the EBNA1 protein (Ambinder et al. 1991; Chen et al. 1993; Frappier and O'Donnell 1991b; Shah et al. 1992; Summers et al. 1996) (Fig. 2.2). In fact, EBNA1 forms extremely stable homodimers which cannot dissociate to monomers without unfolding the protein (Bochkarev et al. 1995). The crystal structure of the DNA binding/dimerisation domain was determined both in solution and bound to the EBNA1 consensus binding site, revealing the mechanisms of homodimerisation and DNA binding (Bochkarev et al. 1996; Bochkarev et al. 1995) (Fig. 3.1). Dimerization is mediated by amino acids 504–604 (referred to as the core domain), which forms an eight-stranded antiparallel β-barrel (four strands from each monomer) with two

L. Frappier, *EBNA1 and Epstein-Barr Virus Associated Tumours*,
SpringerBriefs in Cancer Research, DOI: 10.1007/978-1-4614-6886-8_3,
© The Author(s) 2013

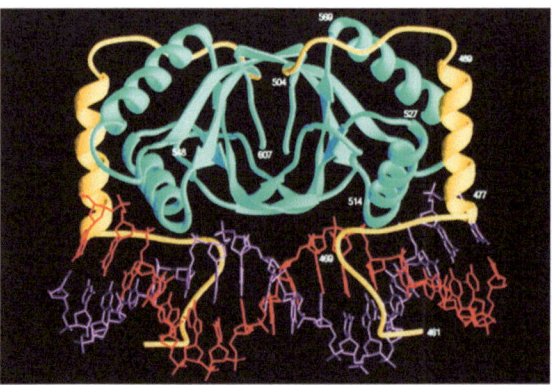

**Fig. 3.1** *Structure of the EBNA1 DNA binding and dimerisation domain.* The structure of the EBNA1 458–607 fragment crystalised on its DNA recognition site is shown along with pertinent amino acid numbers. The core DNA binding/dimerisation domain is shown in blue and the flanking DNA binding domain is shown in yellow. Reprinted with permission from Bochkarev et al. 1996 (Cell vol. 84(5): 791–800)

α-helices per monomer. This core domain is remarkably similar in structure to the DNA binding and dimerisation domain of the E2 protein of bovine papillomavirus, even though the two proteins share no sequence homology (Bochkarev et al. 1995; Edwards et al. 1998; Hegde et al. 1992). The co-crystal structure indicated that the main sequence–specific DNA interactions of EBNA1 occurred through a region just *N*-terminal to the core domain (referred to as the flanking domain), between amino acids 461 and 503. These DNA interactions occurred through an α-helix oriented perpendicular to the DNA and through an extended polypeptide chain that tunnels along the base of the minor groove of the DNA. This flanking domain does not appear to exist in E2, which uses an α-helix in the core domain for sequence–specific interactions with the DNA major groove. The equivalent α-helix in the EBNA1 core domain was not observed to make sequence–specific contacts with the DNA in the EBNA1-DNA co-crystal structure; however, an important role for this helix in DNA binding was identified when point mutations in this helix were found to greatly decrease DNA binding (Cruickshank et al. 2000). The data suggest that the core domain makes the initial contacts with the DNA, followed by loading of the flanking domain onto the DNA to generate the stable EBNA1-DNA complex. In keeping with this model, thermodynamic and kinetic analyses of the EBNA1 DNA binding domain-DNA interaction indicated that DNA association and dissociation is a two-step process (Oddo et al. 2006).

The structure of the EBNA1 DNA binding and dimerisation domain bound to a single recognition site showed that EBNA1 altered the structure of the DNA, so that it was smoothly bent with localized regions of helical overwinding and underwinding (Bochkarev et al. 1996) (Fig. 3.1). The overwinding of the DNA results in a T residue in two of the four recognition sites in the DS being sensitive to oxidation by permanganate (Bochkarev et al. 1998; Frappier and O'Donnell 1992; Hearing et al. 1992; Hsieh et al. 1993; Summers et al. 1997). This DNA distortion is caused by the EBNA1 flanking domain residues that traverse along the minor groove (amino acids 463–468) (Bochkarev et al. 1998; Summers et al. 1997). The assembly of EBNA1 dimers on adjacent sites in the DS occurs cooperatively (Harrison et al. 1994; Summers et al. 1996), and is predicted to involve changes in the DNA structure (such as unwinding), in addition to those observed for EBNA1 bound to a single site, in order to accommodate the closely packed dimers (Bochkarev et al. 1996). A requirement for a precise interaction between these dimers and/or a specific change in DNA structure may be the reason that adjacent EBNA1 binding sites must be 3 bp apart in order to support DNA replication (Bashaw and Yates 2001; Harrison et al. 1994). In addition, the interactions of the multiple DNA-bound EBNA1 dimers within the DS and FR elements likely contribute to the pronounced bending of these elements and to the appearance of EBNA1 as a single complex on each element (Bashaw and Yates 2001; Frappier and O'Donnell 1991a; Goldsmith et al. 1993).

The EBNA1 complexes that assemble at the FR and DS elements of *oriP* have also been observed to interact with each other cause the looping out of the intervening DNA (when interactions occur within an *oriP* molecule; referred to as DNA looping) and the cross-linking of multiple *oriP* molecules (when interactions occur between *oriP* molecules; referred to as DNA linking) (Frappier and O'Donnell 1991a; Goldsmith et al. 1993; Middleton and Sugden 1992; Su et al. 1991). The DNA looping and linking interactions stabilise EBNA1 on the DS and involve homotypic interactions mediated by the central Gly-Arg-rich region (amino acids 325–376). In addition, residues 40–89 can contribute to the looping and linking interactions but to a lesser degree (Avolio-Hunter and Frappier 1998; Frappier et al. 1994; Laine and Frappier 1995; Mackey et al. 1995; Mackey and Sugden 1999). These long-distance interactions of EBNA1 do not require large EBNA1 complexes but can also occur between single DNA-bound EBNA1 (Goldsmith et al. 1993). The contribution of DNA looping and linking to EBNA1 functions is unclear, but the amino acids that mediate these interactions are also important for EBNA1 replication, segregation and transcriptional activation functions (Mackey and Sugden 1999; Shire et al. 1999; Wu et al. 2002).

## 3.2 Interactions with Cellular DNA

The ability of EBNA1 to bind specific EBV DNA sequences raises the possibility that EBNA1 might also recognise specific cellular DNA sequences through its DNA binding domain. If this were the case then EBNA1 might affect the expression of neighbouring cellular genes as it does for EBV genes. This possibility has prompted several laboratories to look for cellular sequences that are bound by EBNA1 and affect cellular gene expression. Genome-wide chromatin immunoprecipitation (ChIP) experiments were performed for EBNA1 in EBV-positive B-cell lines by two groups (Dresang et al. 2009; Lu et al. 2010). Dresang et al. (Dresang et al. 2009) used promoter arrays to identify several EBNA1-associated DNA fragments, some of which were shown to be directly bound by EBNA1 in vitro. From this analysis, an EBNA1 recognition sequence was identified that was distinct from that in *oriP*. However, reporter assays did not detect any effect of EBNA1 binding to this sequence on gene expression, so it is not clear if there is any functional significance to these EBNA1-cellular DNA interactions. Lu et al. (2010) identified many EBNA1-associated DNA fragments by deep sequencing, including a cluster of high-affinity sites on chromosome 11 between the divergent FAM55D and FAM55B genes. Like the findings of Dresang et al. (2009), many of the EBNA1-associated sequences were unrelated to those in *oriP* and several were close to transcriptional start sites for cellular genes. Expression of a subset of these cellular genes was shown to be decreased upon EBNA1 depletion and increased by EBNA1 expression, although it was not clear if this involved direct or indirect association of EBNA1 with the cellular DNA. However, expression of the FAM55D and FAM55B genes was not affected by EBNA1, even though they flank the EBNA1 high affinity sites which could be directly bound by EBNA1 (Lu et al. 2010).

In another study, d'Herouel et al. (2010) used nearest neighbor position weight matrices to identify repeated EBNA1 binding sites in the human genome. The rationale of this approach is that EBNA1 binding to single recognitions sites within the EBV genome has never been found to be sufficient for any EBNA1-associated function, but rather cooperative assembly on multiple tandem sites is required. Therefore, it seems unlikely that a single EBNA1-bound site in the cellular genome would activate transcription. D'Herouel identified putative repeated EBNA1 sites in cellular DNA, some of which corresponded to sites identified and experimentally verified by Dresang et al. (2009). In addition, they identified tandem EBNA1 sites in 40 novel regions of the genome, although the significance of these findings remains to be determined.

Others have tried to identify cellular genes controlled by EBNA1 by using microarray approaches to identify cellular transcripts whose levels change in the presence and absence of EBNA1 (Canaan et al. 2009; Kang et al. 2001; Lu et al. 2011; Wood et al. 2007). The first report was by Kang et al. (2001) in EBV-positive B-cells, in which they overexpressed the EBNA1 DNA binding domain, known to inhibit the EBNA1 transactivation function at the EBV genome by competing for EBNA1 binding sites (Kirchmaier and Sugden 1997). This study did not identify any cellular transcripts that were altered by this dominant-negative EBNA1 fragment, suggesting that endogenous EBNA1 was not contributing to the expression of any cellular genes through specific interactions with its DNA binding domain. Later, Wood et al. (2007) identified STAT1 as being upregulated by EBNA1 expression in carcinoma cell lines, but whether or not this involves EBNA1 interactions with the cellular DNA was not determined. Canaan et al. (2009) identified a small percentage of B-cell and 293 cell transcripts that were affected by EBNA1 and found that EBNA1 ChIP'd to the promoters of most of these genes, suggesting it directly regulated them. They also identified sequence motifs associated with up- and down regulation by EBNA1, all of which were distinct from EBNA1-bound sequences in the EBV genome. However, it was not determined whether EBNA1 bound directly to these sequences or was associated with them through interactions with cellular DNA-bound proteins. Finally,

Lu et al. (2011) compared cell cycle specific transcripts from EBV-negative B-cells with and without EBNA1 expression. They identified several transcripts that were increased by EBNA1 including those coding for survivin, a protein that negatively regulates apoptosis by inhibiting caspase activity. EBNA1 was subsequently shown to induce survivin protein levels and to activate the survivin promoter through Sp1 sites at the promoter (Lu et al. 2011). This suggests that EBNA1 may associate indirectly with the survivin promoter through Sp1. However, it is interesting that the induction of survivin expression by EBNA1 requires the 65–89 transcriptional activation sequence of EBNA1, strongly suggesting that EBNA1 is directly responsible for the transactional activation of the survivin gene (Lu et al. 2011). Therefore, EBNA1 may affect the transcription of specific cellular genes through promoter associations, although there are currently no clear examples of cellular genes that are regulated by direct EBNA1 binding to specific cellular DNA sequences.

# References

Ambinder, R. F., Mullen, M., Chang, Y., Hayward, G. S., & Hayward, S. D. (1991). Functional domains of Epstein-Barr nuclear antigen EBNA-1. *Journal of Virology, 65*, 1466–1478.

Ambinder, R. F., Shah, W. A., Rawlins, D. R., Hayward, G. S., & Hayward, S. D. (1990). Definition of the sequence requirements for binding of the EBNA-1 protein to its palindromic target sites in Epstein-Barr virus DNA. *Journal of Virology, 64*, 2369–2379.

Avolio-Hunter, T. M., & Frappier, L. (1998). Mechanistic studies on the DNA linking activity of the Epstein-Barr nuclear antigen 1. *Nucleic Acids Research, 26*, 4462–4470.

Bashaw, J. M., & Yates, J. L. (2001). Replication from oriP of Epstein-Barr virus requires exact spacing of two bound dimers of EBNA1 which bend DNA. *Journal of Virology, 75*, 10603–10611.

Bochkarev, A., Barwell, J., Pfuetzner, R., Bochkareva, E., Frappier, L., & Edwards, A. M. (1996). Crystal structure of the DNA-binding domain of the Epstein-Barr virus origin binding protein, EBNA1, bound to DNA. *Cell, 84*, 791–800.

Bochkarev, A., Barwell, J., Pfuetzner, R., Furey, W., Edwards, A., & Frappier, L. (1995). Crystal structure of the DNA binding domain of the Epstein-Barr virus origin binding protein EBNA1. *Cell, 83*, 39–46.

Bochkarev, A., Bochkareva, E., Frappier, L., & Edwards, A. M. (1998). 2.2A structure of a permanganate-sensitive DNA site bound by the Epstein-Barr virus origin binding protein, EBNA1. *Journal of Molecular Biology, 284*, 1273–1278.

Canaan, A., Haviv, I., Urban, A. E., Schulz, V. P., Hartman, S., Zhang, Z., et al. (2009). EBNA1 regulates cellular gene expression by binding cellular promoters. *Proceedings of the National Academy of Sciences USA, 106*, 22421–22426.

Chen, M.-R., Middeldorp, J. M., & Hayward, S. D. (1993). Separation of the complex DNA binding domain of EBNA-1 into DNA recognition and dimerization subdomains of novel structure. *Journal of Virology, 67*, 4875–4885.

Cruickshank, J., Davidson, A., Edwards, A. M., & Frappier, L. (2000). Two domains of the Epstein-Barr virus origin DNA binding protein, EBNA1, orchestrate sequence-specific DNA binding. *Journal of Biological Chemistry, 275*, 22273–22277.

d'Herouel, A. F., Birgersdotter, A., & Werner, M. (2010). FR-like EBNA1 binding repeats in the human genome. *Virology, 405*, 524–529.

Dresang, L. R., Vereide, D. T., & Sugden, B. (2009). Identifying sites bound by Epstein-Barr virus nuclear antigen 1 (EBNA1) in the human genome: Defining a position-weighted matrix to predict sites bound by EBNA1 in viral genomes. *Journal of Virology, 83*, 2930–2940.

Edwards, A. M., Bochkarev, A., & Frappier, L. (1998). Origin DNA -binding proteins. *Current Opinion in Structural Biology, 8*, 49–53.

Frappier, L., Goldsmith, K., & Bendell, L. (1994). Stabilization of the EBNA1 protein on the Epstein-Barr virus latent origin of DNA replication by a DNA looping mechanism. *Journal of Biological Chemistry, 269*, 1057–1062.

Frappier, L., & O'Donnell, M. (1991a). Epstein-Barr nuclear antigen 1 mediates a DNA loop within the latent replication origin of Epstein-Barr virus. *Proceedings of the National Academy of Sciences USA, 88*, 10875–10879.

Frappier, L., & O'Donnell, M. (1991b). Overproduction, purification and characterization of EBNA1, the origin binding protein of Epstein-Barr virus. *Journal of Biological Chemistry, 266*, 7819–7826.

Frappier, L., & O'Donnell, M. (1992). EBNA1 distorts *oriP*, the Epstein-Barr virus latent replication origin. *Journal of Virology, 66*, 1786–1790.

Goldsmith, K., Bendell, L., & Frappier, L. (1993). Identification of EBNA1 amino acid sequences required for the interaction of the functional elements of the Epstein-Barr virus latent origin of DNA replication. *Journal of Virology, 67*, 3418–3426.

Harrison, S., Fisenne, K., & Hearing, J. (1994). Sequence requirements of the Epstein-Barr virus latent origin of DNA replication. *Journal of Virology, 68*, 1913–1925.

Hearing, J., Mulhaupt, Y., & Harper, S. (1992). Interaction of Epstein-Barr virus nuclear antigen 1 with the viral latent origin of replication. *Journal of Virology, 66*, 694–705.

Hegde, R. S., Grossman, S. R., Laimins, L. A., & Sigler, P. B. (1992). Crystal structure at 1.7Å of the bovine papillomavirus-1 E2 DNA-binding protein bound to its DNA target. *Nature, 359*, 505–512.

Hsieh, D.-J., Camiolo, S. M., & Yates, J. L. (1993). Constitutive binding of EBNA1 protein to the Epstein-Barr virus replication origin, oriP, with distortion of DNA structure during latent infection. *EMBO Journal, 12*, 4933–4944.

Jones, C. H., Hayward, S. D., & Rawlins, D. R. (1989). Interaction of the lymphocyte-derived Epstein-Barr virus nuclear antigen EBNA-1 with its DNA-binding sites. *Journal of Virology, 63*, 101–110.

Kang, M. S., Hung, S. C., & Kieff, E. (2001). Epstein-Barr virus nuclear antigen 1 activates transcription from episomal but not integrated DNA and does not alter lymphocyte growth. *Proceedings of National Academy of Sciences U S A, 98*, 15233–15238.

Kirchmaier, A. L., & Sugden, B. (1997). Dominant-negative inhibitors of EBNA1 of Epstein-Barr virus. *Journal of Virology, 71*, 1766–1775.

Laine, A., & Frappier, L. (1995). Identification of Epstein-Barr nuclear antigen 1 protein domains that direct interactions at a distance between DNA-bound proteins. *Journal of Biological Chemistry, 270*, 30914–30918.

Lu, F., Wikramasinghe, P., Norseen, J., Tsai, K., Wang, P., Showe, L., et al. (2010). Genome-wide analysis of host-chromosome binding sites for Epstein-Barr Virus Nuclear Antigen 1 (EBNA1). *Virology Journal, 7*, 262.

Lu, J., Murakami, M., Verma, S. C., Cai, Q., Haldar, S., Kaul, R., et al. (2011). Epstein-Barr Virus nuclear antigen 1 (EBNA1) confers resistance to apoptosis in EBV-positive B-lymphoma cells through up-regulation of survivin. *Virology, 410*, 64–75.

Mackey, D., Middleton, T., & Sugden, B. (1995). Multiple regions within EBNA1 can link DNAs. *Journal of Virology, 69*, 6199–6208.

Mackey, D., & Sugden, B. (1999). The linking regions of EBNA1 are essential for its support of replication and transcription. *Molecular and Cellular Biology, 19*, 3349–3359.

Middleton, T., & Sugden, B. (1992). EBNA1 can link the enhancer element to the initiator element of the Epstein-Barr virus plasmid origin of DNA replication. *Journal of Virology, 66*, 489–495.

Nonkwelo, C., Skinner, J., Bell, A., Rickinson, A., & Sample, J. (1996). Transcription start site downstream of the Epstein-Barr virus (EBV) Fp promoter in early-passage Burkitt lymphoma cells define a fourth promoter for expression of the EBV EBNA1 protein. *Journal of Virology, 70*, 623–627.

Oddo, C., Freire, E., Frappier, L., & de Prat-Gay, G. (2006). Mechanism of DNA recognition at a viral replication origin. *Journal of Biological Chemistry, 281*, 26893–26903.

Rawlins, D. R., Milman, G., Hayward, S. D., & Hayward, G. S. (1985). Sequence-specific DNA binding of the Epstein-Barr virus nuclear antigen (EBNA1) to clustered sites in the plasmid maintenance region. *Cell, 42*, 859–868.

Sample, J., Henson, E. B. D., & Sample, C. (1992). The Epstein-Barr virus nuclear protein 1 promoter active in type I latency is autoregulated. *Journal of Virology, 66*, 4654–4661.

Shah, W. A., Ambinder, R. F., Hayward, G. S., & Hayward, S. D. (1992). Binding of EBNA-1 to DNA creates a protease-resistant domain that encompasses the DNA recognition and dimerization functions. *Journal Virology, 66*, 3355–3362.

Shire, K., Ceccarelli, D. F. J., Avolio-Hunter, T. M., & Frappier, L. (1999). EBP2, a human protein that interacts with sequences of the Epstein-Barr nuclear antigen 1 important for plasmid maintenance. *Journal of Virology, 73*, 2587–2595.

Su, W., Middleton, T., Sugden, B., & Echols, H. (1991). DNA looping between the origin of replication of Epstein-Barr virus and its enhancer site: Stabilization of an origin complex with Epstein-Barr nuclear antigen 1. *Proceedings of the National Academy of Sciences USA, 88*, 10870–10874.

Summers, H., Barwell, J. A., Pfuetzner, R. A., Edwards, A. M., & Frappier, L. (1996). Cooperative assembly of EBNA1 on the Epstein-Barr virus latent origin of replication. *Journal of Virology, 70*, 1228–1231.

Summers, H., Fleming, A., & Frappier, L. (1997). Requirements for EBNA1-induced permanganate sensitivity of the Epstein-Barr virus latent origin of DNA replication. *Journal of Biological Chemistry, 272*, 26434–26440.

Wood, V. H., O'Neil, J. D., Wei, W., Stewart, S. E., Dawson, C. W., & Young, L. S. (2007). Epstein-Barr virus-encoded EBNA1 regulates cellular gene transcription and modulates the STAT1 and TGFbeta signaling pathways. *Oncogene, 26*, 4135–4147.

Wu, H., Kapoor, P., & Frappier, L. (2002). Separation of the DNA replication, segregation and transcriptional activation functions of Epstein-Barr nuclear antigen 1. *Journal of Virology, 76*, 2480–2490.

Yoshioka, M., Crum, M. M., & Sample, J. T. (2008). Autorepression of Epstein-Barr virus nuclear antigen 1 expression by inhibition of pre-mRNA processing. *Journal of Virology, 82*, 1679–1687.

# Chapter 4
# EBNA1 Contributions to EBV-Associated Tumours

## 4.1 B Cell Lymphomas

B cells are the main site of EBV latency, and EBV latent infection involves the immortalisation of resting B lymphocytes. Therefore, it is not surprising that most of the tumours associated with EBV infection are B cell lymphomas. Several different types of EBV-associated B cell lymphomas are known, each with distinct profiles of latent gene expression corresponding to different latency types (Thorley-Lawson and Gross 2004). However, EBNA1 is expressed in all of them. In Burkitt's lymphoma, the first known EBV-associated tumour, EBNA1 is the only EBV protein expressed corresponding to latency I (reviewed in Bornkamm (2009)). Hodgkin's disease is another EBV-associated lymphoma, and in this cancer EBNA1 is expressed in combination with the latent membrane proteins LMP1 and LMP2A (latency II) (Deacon et al. 1993; Hummel et al. 1992). EBV is also a causative agent in posttransplant lymphoproliferative disorder (PTLD), an aggressive cancer that can arises upon immunosuppression, such as is induced after organ transplant. These cells are highly immunogenic, and hence normally held in check by the immune system, due to expression of all six of the EBNA proteins in conjunction with LMPs 1, 2A and 3B (latency III). This latency III profile is also what occurs when EBV infects resting B lymphocytes in tissue culture, and hence is the best studied of the B cell latent infections.

EBNA1 is very important for the initial immortalisation of B cells by EBV in tissue culture and is essential for maintaining EBV latent episomes in proliferating cells (Hume et al. 2003). While these features point to an important role for EBNA1 in EBV-associated cancers, it was originally thought that this role was indirect. Namely, by allowing the EBV genomes to persist, EBNA1 enabled the continued expression of other EBV latency proteins, EBV stable RNA molecules (EBERs) and the more recently discovered EBV-encoded micro RNAs, which then induce cell transformation. Indeed, the ability of EBNA1 to transactivate the expression of other EBV latency genes has been shown to be an important part of the mechanism by which EBNA1 contributes to B cell immortalisation, as the

L. Frappier, *EBNA1 and Epstein-Barr Virus Associated Tumours*,
SpringerBriefs in Cancer Research, DOI: 10.1007/978-1-4614-6886-8_4,
© The Author(s) 2013

efficiency of EBV-induced immortalisation is greatly decreased by a mutation that abrogates the transcriptional activation function of EBNA1 (Altmann et al. 2006). However, the fact that EBNA1 is the only EBV protein expressed in Burkitt's lymphoma shows that other EBV proteins are not always required to induce or maintain the transformed state.

There is also some evidence that EBNA1 itself is important for maintaining B cell transformation. For example, Kennedy et al. (2003) showed that a dominant-negative version of EBNA1 that interferes with endogenous EBNA1 function by blocking EBNA1 recognition sites, could induce cell death by apoptosis in several Burkitt's lymphoma cell lines. Since dominant-negative EBNA1 also leads to the loss of EBV eepisome, in most Burkitt's cell lines, the requirement for EBNA1 function in cell survival might simply be due to a need to maintain the expression of RNA molecules (EBERs or miRNA) from the EBV episomes. However, similar effects were also reported for Namalwa cells, which have EBV integrated into the cellular chromosomes, suggesting that the survival effect of EBNA1 is not due to a requirement for EBV episomes (Kennedy et al. 2003). Similarly, a study that used a DNA ligand to interfere with DNA binding by EBNA1 in lymphoblastoid cell lines found that this led to loss of EBV episomes and inhibition of cell proliferation (Yasuda et al. 2011). In addition, EBNA1 silencing in Raji Burkitt's lymphoma cells was shown to decrease cell proliferation (Hong et al. 2006). On the other hand, Kang et al. (2001) examined the effect of dominant–negative EBNA1 on the growth of EBV-positive lymphoblastoid cell lines, and in contrast to the results of Kennedy et al., they did not detect any effect on the growth or survival of these cells. However, dominant–negative EBNA1 would only interfere with EBNA1 functions that require sequence–specific DNA binding through the DNA binding domain, and as discussed in Chap. 5 current data suggests that several cellular effects of EBNA1 are independent of DNA interactions.

Other studies have assessed the contributions of EBNA1 to tumourigenicity in the absence of EBV genomes with varying results. Kube et al. (1999) expressed EBNA1 in the EBV-negative Hodgkin's lymphoma cell line L428 and found that, while these cells were non-tumourigenic in SCID mice, the presence of EBNA1 enhanced the ability of these cells to form lymphomas in nonobese diabetic-SCID mice. Wilson et al. (1996) first reported that EBNA1 expression in the B cell compartment in two different C57BL/6 mouse lines resulted in a high incidence of B cell lymphomas. Subsequently, lymphoid cells from these EBNA1 transgenic mice were shown to have increased growth or survival capacity in culture (Tsimbouri et al. 2002). In addition, when EBNA1 and Myc transgenic mice were cross-bred, synergy in lymphomagenesis was observed between Myc and EBNA1 in some crosses, resulting in a reduced tumour latency period (Drotar et al. 2003). In contrast to these studies, Kang et al. (2005, 2008) developed several FVB and C57BL/6 mouse lines expressing EBNA1 in B-lymphocytes and did not observe any increase in tumour incidence due to EBNA1 expression. Therefore, although there is disagreement on whether or not EBNA1 expression is sufficient to induce tumours, as a whole the results seem to suggest that EBNA1 can contribute to B cell tumourigenicity under certain circumstances.

## 4.2 Nasopharyngeal Carcinoma

EBV is widely recognised as a causative agent of nasopharyngeal carcinoma (NPC), as these tumours are clonal expansions of EBV-positive cells, in contrast to the surrounding normal epithelium which is EBV-negative (reviewed in Raab-Traub (2002); Tao and Chan (2007)). The NPC cells consistently express EBNA1 and a secreted EBV protein called BARF1, and occasionally express LMP1. The EBV-encoded EBERs and a variety of micro RNAs are also expressed. Since EBNA1 is the only EBV protein consistently present inside the NPC cells, it may play an important role in promoting or maintaining cell proliferation or survival. In keeping with this hypothesis, down-regulation of EBNA1 in NPC-derived cells with siRNA resulted in decreased cell proliferation (Yin and Flemington 2006). Conversely, Sheu et al. (1996) found that EBNA1 expression in the EBV-negative HONE-1 NPC cells resulted in less differentiated and more rapidly growing tumours in nude mice as well as increased metastatic capabilities. Consistent with these results, Kaul et al. (2007) found that expression of EBNA1 in a breast carcinoma cell line promoted the rate of tumour growth in nude mice, increased the percentage of mice that got tumours and caused a 3-fold increase in lung metastases. The effect on metastasis is consistent with a previous report from this group that EBNA1 can bind Nm23-H1 and relocalising it to the nucleus, thereby counteracting the ability of Nm23-H1 to suppress cell migration (Murakami et al. 2005).

The connection between EBNA1 and metastases was further strengthened by a recent 2D Difference Gel Electrophoresis (DiGE) study, in which the nuclear proteomes of CNE2 NPC cells with and without EBNA1 expression were compared. The presence of EBNA1 was found to increase the nuclear levels of three proteins involved in metastases, stathmin 1, maspin, and Nm23-H1. This confirmed that the observations of Murakami et al. (2005) are also relevant for NPC cells and suggested that EBNA1 affects the metastatic potential of NPC cells in multiple ways (Cao et al. 2011). In addition, a microarray study on NPC cells found that EBNA1 increased VEGF expression, suggesting that EBNA1 may increase angiogenesis, which could contribute to NPC development and metastasis (O'Neil et al. 2008). Finally, as discussed in detail in Chap. 5, in the context of NPC cells, EBNA1 has been found to down-regulate p53, interfere with DNA repair, inhibit apoptosis and promote cell survival, at least in part through the disruption of PML nuclear bodies (Sivachandran et al. 2008). Therefore, there is considerable evidence that EBNA1 contributes to NPC.

While EBNA1 is able to replicate and segregate *oriP* plasmids in NPC and other epithelial cell lines, it should be pointed out that EBV episomes are typically lost from NPC cells once they are grown in culture. The fact that B cell stably maintains EBV episomes and epithelial cells do not is thought to be due to differences in the efficiency of EBNA1-mediated mitotic segregation. The molecular basis for this difference is not known but two studies have indicated that maintaining the appropriate level of EBNA1 expression for optimal segregation in NPC cells may be part of the problem (Shannon-Lowe et al. 2009; Sivachandran et al. 2011).

## 4.3  Gastric Carcinoma

Another epithelial tumour that is associated with EBV infection is gastric carcinoma, particularly those tumours at the top of the stomach. EBV-positive gastric tumours are estimated to account for 10 % of all gastric carcinomas worldwide and display a latency I-like profile in which EBNA1 and LMP2A and BARF1 are the only EBV proteins expressed (Chen et al. 2012; Fukayama 2010; Tsao et al. 2012). So far, studies on EBNA1 contributions to gastric carcinoma are limited but results are similar to those in NPC cells. For example, like in NPC cells, EBNA1 expression in gastric carcinoma cells was reported to enhance tumourigenicity in nude mice (Cheng et al. 2010). In addition, as in NPC cells, EBNA1 induces the loss of PML nuclear bodies in AGS gastric carcinoma cells, resulting in impaired p53 function, decreased apoptosis in response to DNA damage and increased cell survival after DNA damage (Sivachandran et al. 2012). The EBNA1-induced loss of PML nuclear bodies observed in cultured cells was confirmed in gastric tumours, where EBV-positive tumours were found to have considerably less PML staining than their EBV-negative counterparts (Sivachandran et al. 2012). Therefore, it appears that EBNA1 expression leads to loss of the tumour-suppressing PML nuclear bodies in EBV-positive gastric tumours which would contribute to oncogenesis or tumour progression.

## References

Altmann, M., Pich, D., Ruiss, R., Wang, J., Sugden, B., & Hammerschmidt, W. (2006). Transcriptional activation by EBV nuclear antigen 1 is essential for the expression of EBV's transforming genes. *Proceedings of the National Academy of Sciences of the United States of America, 103*, 14188–14193.

Bornkamm, G. W. (2009). Epstein-Barr virus and its role in the pathogenesis of Burkitt's lymphoma: an unresolved issue. *Seminars in Cancer Biology, 19*, 351–365.

Cao, J. Y., Mansouri, S., & Frappier, L. (2011). Changes in the nasopharyngeal carcinoma nuclear proteome induced by the EBNA1 protein of Epstein-Barr virus reveal potential roles for EBNA1 in metastasis and oxidative stress responses. Journal of Virology, *85*(19), 10425–10430.

Chen, J. N., He, D., Tang, F., & Shao, C. K. (2012). Epstein-Barr virus-associated gastric carcinoma: a newly defined entity. *Journal of Clinical Gastroenterology, 46*, 262–271.

Cheng, T. C., Hsieh, S. S., Hsu, W. L., Chen, Y. F., Ho, H. H., & Sheu, L. F. (2010). Expression of Epstein-Barr nuclear antigen 1 in gastric carcinoma cells is associated with enhanced tumorigenicity and reduced cisplatin sensitivity. *International Journal of Oncology, 36*, 151–160.

Deacon, E. M., Pallesen, G., Niedobitek, G., Crocker, J., Brooks, L., Rickinson, A. B., et al. (1993). Epstein-Barr virus and Hodgkin's disease: transcriptional analysis of virus latency in the malignant cells. *Journal of Experimental Medicine, 177*, 339–349.

Drotar, M. E., Silva, S., Barone, E., Campbell, D., Tsimbouri, P., Jurvansu, J., et al. (2003). Epstein-Barr virus nuclear antigen-1 and Myc cooperate in lymphomagenesis. *International Journal of Cancer, 106*, 388–395.

Fukayama, M. (2010). Epstein-Barr virus and gastric carcinoma. *Pathology International, 60*, 337–350.

Hong, M., Murai, Y., Kutsuna, T., Takahashi, H., Nomoto, K., Cheng, C. M., et al. (2006). Suppression of Epstein-Barr nuclear antigen 1 (EBNA1) by RNA interference inhibits proliferation of

EBV-positive Burkitt's lymphoma cells. *Journal of Cancer Research and Clinical Oncology, 132*, 1–8.

Hume, S., Reisbach, G., Feederle, R., Delecluse, H.-J., Bousset, K., Hammerschmidt, W., et al. (2003). The EBV nuclear antigen 1 (EBNA1) enhances B cell immortalization several thousandfold. *Proceedings of the National Academy of Sciences, 100*, 10989–10994.

Hummel, M., Anagnostopoulos, I., Dallenbach, F., Korbjuhn, P., Dimmler, C., & Stein, H. (1992). EBV infection patterns in Hodgkin's disease and normal lymphoid tissue: expression and cellular localization of EBV gene products. *British Journal of Haematology, 82*, 689–694.

Kang, M. S., Hung, S. C., & Kieff, E. (2001). Epstein-Barr virus nuclear antigen 1 activates transcription from episomal but not integrated DNA and does not alter lymphocyte growth. *Proceedings of the National Academy of Sciences of the United States of America, 98*, 15233–15238.

Kang, M. S., Lu, H., Yasui, T., Sharpe, A., Warren, H., Cahir-McFarland, E., et al. (2005). Epstein-Barr virus nuclear antigen 1 does not induce lymphoma in transgenic FVB mice. *Proceedings of the National Academy of Sciences of the United States of America, 102*, 820–825.

Kang, M. S., Soni, V., Bronson, R., & Kieff, E. (2008). Epstein-Barr virus nuclear antigen 1 does not cause lymphoma in c57bl/6 j mice. *Journal of Virology, 82*, 4180–4183.

Kaul, R., Murakami, M., Choudhuri, T., & Robertson, E. S. (2007). Epstein-Barr virus latent nuclear antigens can induce metastasis in a nude mouse model. *Journal of Virology, 81*, 10352–10361.

Kennedy, G., Komano, J., & Sugden, B. (2003). Epstein-Barr virus provide a survival factor to Burkitt's lymphomas. *Proceedings of the National Academy of Sciences, 100*, 14269–14274.

Kube, D., Vockerodt, M., Weber, O., Hell, K., Wolf, J., Haier, B., et al. (1999). Expression of Epstein-Barr virus nuclear antigen 1 is associated with enhanced expression of CD25 in the hodgkin cell line L428. *Journal of Virology, 73*, 1630–1636.

Murakami, M., Lan, K., Subramanian, C., & Robertson, E. S. (2005). Epstein-Barr virus nuclear antigen 1 interacts with Nm23-H1 in lymphoblastoid cell lines and inhibits its ability to suppress cell migration. *Journal of Virology, 79*, 1559–1568.

O'Neil, J. D., Owen, T. J., Wood, V. H., Date, K. L., Valentine, R., Chukwuma, M. B., et al. (2008). Epstein-Barr virus-encoded EBNA1 modulates the AP-1 transcription factor pathway in nasopharyngeal carcinoma cells and enhances angiogenesis in vitro. *Journal of General Virology, 89*, 2833–2842.

Raab-Traub, N. (2002). Epstein-Barr virus in the pathogenesis of NPC. *Seminars in Cancer Biology, 12*, 431–441.

Shannon-Lowe, C., Adland, E., Bell, A. I., Delecluse, H. J., Rickinson, A. B., & Rowe, M. (2009). Features distinguishing Epstein-Barr virus infections of epithelial cells and B cells: viral genome expression, genome maintenance, and genome amplification. *Journal of Virology, 83*, 7749–7760.

Sheu, L. F., Chen, A., Meng, C. L., Ho, K. C., Lee, W. H., Leu, F. J., et al. (1996). Enhanced malignant progression of nasopharyngeal carcinoma cells mediated by the expression of Epstein-Barr nuclear antigen 1 in vivo. *Journal of Pathology, 180*, 243–248.

Sivachandran, N., Sarkari, F., & Frappier, L. (2008). Epstein-Barr nuclear antigen 1 contributes to nasopharyngeal carcinoma through disruption of PML nuclear bodies. *PLoS Pathogens, 4*, e1000170.

Sivachandran, N., Thawe, N. N., & Frappier, L. (2011). Epstein-Barr virus nuclear antigen 1 replication and segregation functions in nasopharyngeal carcinoma cell lines. *Journal of Virology, 85*, 10425–10430.

Sivachandran, N., Dawson, C. W., Young, L. S., Liu, F. F., Middeldorp, J., & Frappier, L. (2012). Contributions of the Epstein-Barr virus EBNA1 protein to gastric carcinoma. *Journal of Virology, 86*, 60–68.

Tao, Q., & Chan, A. T. (2007). Nasopharyngeal carcinoma: molecular pathogenesis and therapeutic developments. *Expert Reviews in Molecular Medicine, 9*, 1–24.

Thorley-Lawson, D. A., & Gross, A. (2004). Persistence of the Epstein-Barr virus and the origins of associated lymphomas. *New England Journal of Medicine, 350,* 1328–1337.

Tsao, S. W., Tsang, C. M., Pang, P. S., Zhang, G., Chen, H., & Lo, K. W. (2012). The biology of EBV infection in human epithelial cells. *Seminars in Cancer Biology, 22,* 137–143.

Tsimbouri, P., Drotar, M. E., Coy, J. L., & Wilson, J. B. (2002). Bcl-xL and RAG genes are induced and the response to IL-2 enhanced in EmuEBNA-1 transgenic mouse lymphocytes. *Oncogene, 21,* 5182–5187.

Wilson, J. B., Bell, J. L., & Levine, A. J. (1996). Expression of Epstein-Barr virus nuclear antigen-1 induces B cell neoplasia in transgenic mice. *EMBO Journal, 15,* 3117–3126.

Yasuda, A., Noguchi, K., Minoshima, M., Kashiwazaki, G., Kanda, T., Katayama, K., et al. (2011). DNA ligand designed to antagonize EBNA1 represses Epstein-Barr virus-induced immortalization. *Cancer Science, 102,* 2221–2230.

Yin, Q., & Flemington, E. K. (2006). siRNAs against the Epstein Barr virus latency replication factor, EBNA1, inhibit its function and growth of EBV-dependent tumor cells. *Virology, 346,* 385–393.

# Chapter 5
# Cellular Effects of EBNA1

EBNA1 functions were initially thought to be restricted to effects on the replication, maintenance and expression of the EBV genome; however, numerous findings in the last decade have identified additional roles for EBNA1 in affecting the cellular environment. Some of these EBNA1 effects may involve direct or indirect interactions with cellular DNA sequences resulting in transcriptional changes (as discussed above), while others involve interactions with cellular proteins. Since EBNA1 can also interact with RNA, there is also the possibility that cellular protein expression could be affected through EBNA1 effects on RNA processing or translation or microRNAs. Although these possibilities have been largely unexplored, an effect of EBNA1 on miR-127 microRNA was recently reported (Onnis et al. 2012). The cellular effects of EBNA1 are summarised below and in Fig. 5.1.

## 5.1 p53 Regulation

DNA tumour viruses commonly down-regulate p53 as part of the mechanism by which they promote cell survival and proliferation, so it was expected that EBV would also have a mechanism of interfering with p53 function in its various forms of latency. A link between EBNA1 and p53 was revealed from proteomic experiments (affinity column profiling and TAP-tagging) in which EBNA1 was found to bind stably to ubiquitin specific protease 7 (USP7 or HAUSP) (Holowaty et al. 2003b). About this time, USP7 was shown to bind p53 and Mdm2 and to stabilise these proteins by removing their polyubiquitin chains that normally trigger degradation (Cummins et al. 2004; Li et al. 2002, 2004). EBNA1, p53, and Mdm2 were all shown to interact with the same binding pocket in the N-terminal TRAF domain of USP7, but EBNA1 did so with higher affinity than p53 or Mdm2 (Holowaty et al. 2003a; Hu et al. 2006; Saridakis et al. 2005; Sheng et al. 2006) (Fig. 5.2). As a result, EBNA1 out-competed p53 and Mdm2 in vitro, inhibiting their interaction with USP7. Co-crystal structures

L. Frappier, *EBNA1 and Epstein-Barr Virus Associated Tumours*,
SpringerBriefs in Cancer Research, DOI: 10.1007/978-1-4614-6886-8_5,
© The Author(s) 2013

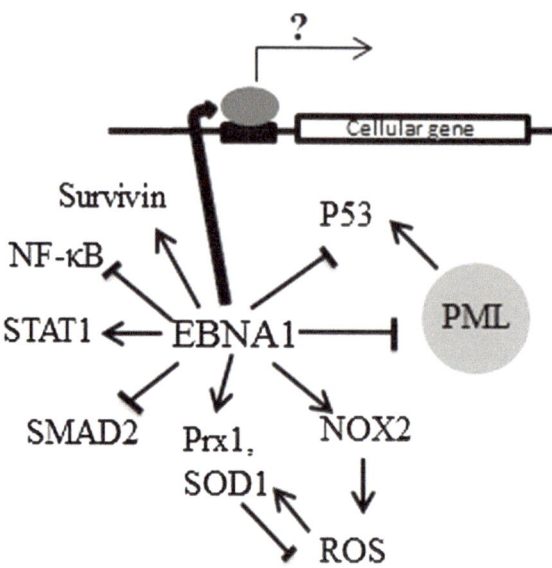

**Fig. 5.1** *Summary of the Cellular Effects of EBNA1.* The cellular proteins whose functions are affected by EBNA1 in ways that might contribute to tumourigenesis are shown, where arrows represent up-regulation and blunt lines represent down-regulation. The possibility that EBNA1 affects cellular genes through interactions with cellular promoter regions is depicted on the top. Reprinted with permission from Frappier 2012

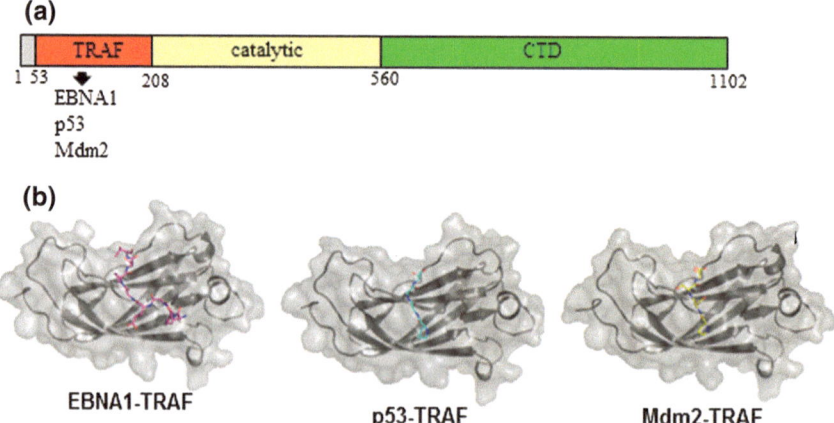

**Fig. 5.2** *USP7.* **a** The organization of USP7 showing the N-terminal TRAF, catalytic and C-terminal (CTD) domains and corresponding amino acid numbers. **b** Structures of the USP7 TRAF domain (*grey*) bound to peptides from EBNA1 (*pink*), p53 (*blue*) and Mdm2 (*yellow*) as determined by crystallography in Saridakis et al. 2005 and Sheng et al. 2006

of these interactions showed that the higher affinity of EBNA1 for USP7 was due to additional contacts with the USP7 TRAF domain, outside of the binding pocket contacted by p53 and Mdm2 (Hu et al. 2006; Saridakis et al. 2005; Sheng et al. 2006).

The EBNA1 amino acids that bind USP7 are 442–448, just N-terminal to the EBNA1 DNA binding and dimerisation domain, and S447 is critical for this interaction (Saridakis et al. 2005) (Fig. 2.2).

Consistent with the in vitro results, EBNA1 expression in U2OS cells was shown to reduce the amount of p53 that accumulated in response to DNA damage, while a USP7-binding mutant of EBNA1 did not have this effect (Saridakis et al. 2005). Similarly, the presence of EBNA1 in CNE2 NPC cells resulted in lower p53 levels after DNA damage (Sivachandran et al. 2008) and the presence of EBNA1 or EBV in AGS or SCM1 gastric carcinoma cells resulted in lower steady-state levels of p53 and less acetylated p53 after DNA damage (Cheng et al. 2010; Sivachandran et al. 2012a). Consistent with these p53 effects, the presence of EBNA1 in all of these cell lines resulted in decreased apoptosis after DNA damage, although this may not be entirely due to the lowered p53 levels. As discussed below, EBNA1 also induces the loss of PML nuclear bodies which are important for apoptosis, in part because they facilitate the activation of p53 through acetylation. Therefore, EBNA1 likely inhibits p53 function in multiple ways (at least in epithelial cells) thereby promoting cell survival. It should be noted that a study on latency III B cell infection concluded that EBV does not downregulate p53 levels or functions (O'Nions et al. 2006). However, it is not clear whether this result means that EBNA1 does not downregulate p53 in lymphoblastoid cells or that another EBV factor expressed in latency III induces p53 levels and that this is counteracted by EBNA1, resulting in unchanged p53 levels.

The mechanism by which EBNA1 lowers p53 levels by binding USP7 is an interesting parallel to the mechanism by which the E6 protein of human papillomavirus lowers p53. In both cases, p53 levels are lowered by increasing its polyubiquitylation. E6 does this by recruiting an E3 ubiqutin ligase (E6AP) to p53 (Howley 2006), while EBNA1 prevents the deubiquitylation of p53 by interfering with the USP7-p53 interaction.

Based on the in vitro and structural data, EBNA1 could also destabilise Mdm2 by interfering with its interaction with USP7, which would result in the stabilisation of p53. To date, there are no reports of EBNA1 decreasing Mdm2 or increasing p53 levels but it remains a possibility that could occur under some circumstances. In addition, several other USP7 targets have been more recently reported, some of which bind USP7 through the N-terminal TRAF domain like EBNA1 (Frappier and Verrijzer 2011; Nicholson and Suresh Kumar 2011). It remains to be determined whether EBNA1 destabilises any of these proteins by blocking their interaction with USP7.

## 5.2  Disruption of PML Nuclear Bodies

Another effect that EBNA1 has been found to have on cells is in inducing the loss of promyelocytic leukaemia (PML) nuclear bodies (also called ND10s). PML bodies are nuclear foci that are important for several cellular processes, including apoptosis, DNA repair and senescence, and accordingly their loss has been associated with the development and/or progression of several tumours (Bernardi and Pandolfi 2007; Guo et al. 2000; Gurrieri et al. 2004; Pearson et al. 2000; Salomoni et al. 2008; Takahashi et al. 2004; Wang et al. 1998). In addition, PML nuclear bodies are part of the innate immune response and act to suppress productive infections of several viruses through inhibition of viral gene expression and replication (Everett and Chelbi-Alix 2007; Geoffroy and Chelbi-Alix 2011; Reichelt et al. 2011). PML proteins, which are comprised of six nuclear isoforms, interact with each other to form the structural basis of the PML nuclear body with which many additional proteins then interact. To counteract the inhibition of viral infection by PML nuclear bodies, many viruses encode proteins that disrupt them, either by inducing the degradation of the PML proteins or by interfering with the interactions of PML proteins to form the bodies (Everett 2001).

EBV latent infection or EBNA1 expression on its own were both found to induce the loss of PML nuclear bodies in NPC and gastric carcinoma cells (Sivachandran et al. 2008, 2012a). A proportion of the EBNA1 was observed to be associated with the PML bodies and was found to preferentially interact with PML isoform IV over the other five nuclear PML isoforms (Sivachandran et al. 2008, 2012b). Subsequent studies showed that EBNA1 induces the degradation of the PML proteins and does so through interactions with both USP7 and casein kinase 2 (CK2), which it recruits to the PML nuclear bodies (Sivachandran et al. 2008, 2010). CK2 is a known regulator of PML in that it phosphorylates PML proteins at a site that primes the PML for poyubiquitylation and degradation (Scaglioni et al. 2006, 2008). EBNA1 interacts with the $\beta$ regulatory subunit of CK2 through EBNA1 amino acids 387–394 (Fig. 2.2) and increases CK2-mediated phosphorylation of PML, seemingly by increasing the association of CK2 with the PML nuclear bodies or proteins (Sivachandran et al. 2010). Since CK2 also regulates many other cellular processes and has several links to oncogenesis, it is possible that the interaction of EBNA1with CK2 affects other CK2-regulated pathways as well but this remains to be determined.

The involvement of USP7 in the EBNA1-mediated disruption of PML nuclear bodies was discovered when an EBNA1 mutant defective in USP7 binding was shown to be unable to disrupt PML bodies or induce the degradation of PML proteins (Sivachandran et al. 2008). Similarly, wild-type EBNA1 did not disrupt PML bodies or induce PML degradation when USP7 was silenced (Sivachandran et al. 2008). These results implicated USP7 as a negative regulator of PML proteins and nuclear bodies. It was subsequently shown that, in the absence of EBV or EBNA1, USP7 silencing increased PML proteins and nuclear bodies, while USP7 overexpression induced their loss (Sarkari et al. 2011). The ability of USP7 to induce PML degradation was independent of the ubiquitin cleavage activity of USP7, as a catalytically inactive USP7 also efficiently induced PML loss, as did the N- and C-terminal protein interaction domains of USP7 (Sarkari et al. 2011). Finally, like EBNA1, USP7 was shown to associate with PML nuclear bodies, largely through interactions with the PML IV isoform (Sarkari et al. 2011).

Since PML nuclear bodies have been shown to be important for apoptosis, DNA repair and p53 acetylation, the effect of EBNA1 on these processes was examined. In keeping with the loss of PML bodies, EBNA1 was found to decrease p53 acetylation, apoptosis and DNA repair, in response to etoposide treatment (Sivachandran et al. 2008, 2012a). The results suggest that cells expressing EBNA1 are more likely to survive with DNA damage, which might contribute to the development of carcinomas. What proportion of these outcomes are due to PML loss as opposed to 53 stabilisation (as discussed above) are unclear. However, it is important to note that the effect of EBV latent infection on PML nuclear bodies has been verified in the context of gastric carcinoma. A comparison of PML staining in EBV-positive and EBV-negative gastric carcinoma samples showed that PML levels were greatly reduced in the presence of EBV, presumably as a result of the action of EBNA1 (Sivachandran et al. 2012a).

Since PML nuclear bodies are known to suppress the lytic infection of at least some herpesviruses (best studied in herpes simplex type 1) and EBNA1 is also expressed in lytic infection (Brink et al. 2001; Lear et al. 1992; Schaefer et al. 1995), another possible outcome of PML loss by EBNA1 is promoting lytic infection. This possibility was recently investigated in EBV-positive AGS gastric carcinoma cells (Sivachandran et al. 2012b). Upon chemical reactivation of EBV to the lytic cycle, EBNA1 silencing was found to inhibit lytic gene expression and viral genome amplification indicating that EBNA1 positively contributes to lytic infection. However, this role in promoting the lytic cycle was not seen when the PML proteins were silenced, suggesting that the role of EBNA1 in lytic infection is in overcoming suppression by PML proteins. In keeping with this interpretation, PML proteins were shown to be suppressive for lytic EBV infection, as PML silencing or disruption of PML nuclear bodies by arsenic trioxide induced EBV reactivation (Sides et al. 2011; Sivachandran et al. 2012b). Also the introduction of individual PML isoforms in cells lacking PML was found to suppress EBV lytic protein expression (Sivachandran et al. 2012b). To date, the effect of EBNA1 on PML bodies has only been observed in epithelial cells, which are the main site of EBV lytic infection. Therefore, it is possible that the main purpose of EBNA1-mediated PML disruption is to promote lytic infection.

## 5.3 Alteration of Signalling

Several studies have indicated that EBNA1 can affect signalling pathways in several ways that might also impact on tumourigenesis. Initially, Wood et al. (2007) compared transcripts in Ad/AH carcinoma cells with and without stable EBNA1 expression and found that the EBNA1-expressing cells had increased expression of STAT1, a protein whose disregulation has several pathological effects including tumourigenesis (Kim and Lee 2007). Similarly, EBNA1 was found to increase STAT1 expression in HONE-1 NPC and AGS gastric carcinoma cells and was further shown to enhance STAT1 phosphorylation and nuclear localisation in response to IFNγ (Wood et al. 2007).

A second finding of the Wood et al. study (2007) was that the presence of EBNA1 decreased the expression of some TGF-β1-responsive genes, consistent with interference with TGF-β signalling. TGF-β1 signalling was then confirmed to be negatively regulated by EBNA1, and EBNA1was shown to lower the levels of SMAD2 complexes needed for TGF-β1-induced transcription, by increasing SMAD2 turnover (Wood et al. 2007). Similar effects of EBNA1 on SMAD2 in Hodgkin's lymphoma cells were also reported, resulting in down-regulation of the protein tyrosine phosphatase receptor kappa (Flavell et al. 2008).

In another study, the effect of EBNA1 on NF-κB reporter plasmids was examined in Ad/AH, AGS and HONE1 carcinoma cell lines, and EBNA1 was found to inhibit NF-κB activity and its DNA interaction (Valentine et al. 2010). Additional studies indicated that EBNA1 decreased the phosphorylation and nuclear translocation of the p65 NF-κB subunit, resulting in a reduction in the amount of p65 in NF-κB nuclear complexes, presumably due to decreased phosphorylation of the p65 kinase, IKKα/β (Valentine et al. 2010). The effect of EBNA1 on NF-κB activity required the EBNA1 61–83 and 325–376 regions associated with transcriptional activation but did not require binding to USP7. Interestingly, p65 localisation in NPC tumour cells is also cytoplasmic, suggesting that EBNA1 affects NF-κB signalling in the context of NPC (Valentine et al. 2010). However, there is currently no evidence that EBNA1 physical interacts with p65, IKKα/β, STAT1 or SMAD2, so the mechanisms by which EBNA1 elicits any of the above effects is unclear and may be indirect due to effects on other cellular proteins.

## 5.4  Induction of Oxidative Stress

EBV infection has been reported to result in increased oxidative stress (Cerimele et al. 2005; Lassoued et al. 2008) and a number of observations suggest that EBNA1 may be at least partly responsible for this. Initially, Gruhne et al. (2009a, b) observed that stable or transient EBNA1 expression in B cell lines increased reactive oxygen (ROS) levels and induced DNA damage foci that were reversed by antioxidants. Additional studies showed that the level of the NOX2 NADPH oxidase was increased in EBV-positive cells and in BJAB cells stably or inducibly expressing EBNA1, and that EBNA1 could increase expression from the NOX2 promoter, although it was not clear if this is a direct effect (Gruhne et al. 2009a). The EBNA1 effect on NOX2 might account for the ROS induction, which could then promote DNA damage resulting in genome instability. It has also been reported that EBNA1 can promote telomere dysfunction which might also be tied to the ROS effect (Kamranvar and Masucci 2011).

Similar conclusions on the effect of EBNA1 on oxidative stress were reached in a proteomics study comparing the nuclear composition of NPC cells with and without EBNA expression (Cao et al. 2011). In this study, EBNA1 was found to increase the levels of several oxidative stress response proteins, including the antioxidants superoxide dismutase 1 (SOD1) and peroxiredoxin 1 (Prx1), which are induced by ROS (Cao et al. 2011). However, transcripts of SOD1 and Prx1 were not increased by EBNA1 suggesting that the upregulation occured at the protein level. Stable EBNA1 expression in the NPC cells was also shown to increase ROS levels and NOX1 and NOX2 proteins and transcripts (Cao et al. 2011). Therefore, EBNA1 appears to affect oxidative stress responses in multiple ways.

## 5.5 Induction of Survivin

Recently, a connection was reported between EBNA1 and survivin, an inhibitor of apoptosis expressed in many cancer cells. By comparing the levels of cellular transcripts in B cells with and without EBNA1 using a cell cycle specific array, Lu et al. (2011) found that EBNA1 increased the levels of survivin transcripts. Survivin protein levels were also increased and EBNA1 was shown to associate with the survivin promoter in B lymphocytes. EBNA1 was not found to bind directly to this promoter but rather appeared to interact with it through the host Sp1 protein. The induction of survivin transcripts and protein was shown to require the EBNA1 65–89 transcriptional activation sequence, suggesting that EBNA1 was transactivating the expression of the survivin gene. Such an EBNA1-induced increase in survivin would be expected to increase the survival of EBV-positive tumour cells by inhibiting apoptosis, and therefore may contribute to EBV-induced lymphomas.

## 5.6 Modulation of NM23-H1

Murakami et al. (2005) detected an interaction between EBNA1 and Nm23-H1, a protein known to suppress metastases and cell migration. EBNA1 co-immunoprecipitated with Nm23-H1 from lymphoblastoid cell lines and shifted its localisation from the cytoplasm to the nucleus. This finding was also supported by a comparative proteomics study by Cao et al. (2011), in which EBNA1 expression in NPC cells was found to increase the nuclear levels of Nm23-H1. EBNA1 was also shown to counteract the inhibitory effect of Nm23-H1 on cell migration, resulting in increased cell motility (Murakami et al. 2005), and in keeping with these results, EBNA1 was later shown to counteract Nm23-H1-mediate suppression of metastases in a nude mouse model (Kaul et al. 2007). The ability of EBNA1 to counteract the cell migration effect of Nm23-H1 was abrogated when the EBNA1 65–89 transactivation sequence was deleted, suggesting that this region might interact with Nm23-H1 (Murakami et al. 2005); however, it is not clear whether or not the EBNA1-Nm23-HI interaction is direct. Interestingly, a pathway-specific microarray study has suggested additional functions for Nm23-H1 in inducing apoptosis and inhibiting cell proliferation (Choudhuri et al. 2010), raising the possibility that EBNA1-induced relocalisation of Nm23-H1 may also be a mechanism to decrease apoptosis and promote cell proliferation.

# References

Bernardi, R., & Pandolfi, P. P. (2007). Structure, dynamics and functions of promyelocytic leukaemia nuclear bodies. *Nature Reviews Molecular Cell Biology, 8*, 1006–1016.

Brink, A. A., Meijer, C. J., Nicholls, J. M., Middeldorp, J. M., & van den Brule, A. J. (2001). Activity of the EBNA1 promoter associated with lytic replication (Fp) in Epstein-Barr virus associated disorders. *Molecular Pathology, 54*, 98–102.

Cao, J.Y., Mansouri, S., and Frappier, L. (2011). Changes in the Nasopharyngeal Carcinoma Nuclear Proteome Induced by the EBNA1 Protein of Epstein-Barr Virus Reveal Potential Roles for EBNA1 in Metastasis and Oxidative Stress Responses. J Virol.

Cerimele, F., Battle, T., Lynch, R., Frank, D. A., Murad, E., Cohen, C., et al. (2005). Reactive oxygen signaling and MAPK activation distinguish Epstein-Barr Virus (EBV)-positive versus EBV-negative Burkitt's lymphoma. *Proceedings of the National Academy of Sciences of the United States of America, 102*, 175–179.

Cheng, T. C., Hsieh, S. S., Hsu, W. L., Chen, Y. F., Ho, H. H., & Sheu, L. F. (2010). Expression of Epstein-Barr nuclear antigen 1 in gastric carcinoma cells is associated with enhanced tumorigenicity and reduced cisplatin sensitivity. *International Journal of Oncology, 36*, 151–160.

Choudhuri, T., Murakami, M., Kaul, R., Sahu, S. K., Mohanty, S., Verma, S. C., et al. (2010). Nm23-H1 can induce cell cycle arrest and apoptosis in B cells. *Cancer Biology and Therapy, 9*, 1065–1078.

Cummins, J. M., Rago, C., Kohli, M., Kinzler, K. W., Lengauer, C., & Vogelstein, B. (2004). Tumour suppression: Disruption of HAUSP gene stabilizes p53. *Nature, 428*, 486–487.

Everett, R. D. (2001). DNA viruses and viral proteins that interact with PML nuclear bodies. *Oncogene, 20*, 7266–7273.

Everett, R. D., & Chelbi-Alix, M. K. (2007). PML and PML nuclear bodies: Implications in antiviral defence. *Biochimie, 89*, 819–830.

Flavell, J. R., Baumforth, K. R., Wood, V. H., Davies, G. L., Wei, W., Reynolds, G. M., et al. (2008). Down-regulation of the TGF-beta target gene, PTPRK, by the Epstein-Barr virus encoded EBNA1 contributes to the growth and survival of Hodgkin lymphoma cells. *Blood, 111*, 292–301.

Frappier, L. (2012). *Seminars in Cancer Biology, 22*(2), 154–161.

Frappier, L., & Verrijzer, C. P. (2011). Gene expression control by protein deubiquitinases. *Current Opinion in Genetics and Development, 21*, 207–213.

Geoffroy, M. C., & Chelbi-Alix, M. K. (2011). Role of promyelocytic leukemia protein in host antiviral defense. *Journal of Interferon and Cytokine Research, 31*, 145–158.

Gruhne, B., Sompallae, R., Marescotti, D., Kamranvar, S. A., Gastaldello, S., & Masucci, M. G. (2009a). The Epstein-Barr virus nuclear antigen-1 promotes genomic instability via induction of reactive oxygen species. *Proceedings of the National Academy of Sciences of the United States of America, 106*, 2313–2318.

Gruhne, B., Sompallae, R., & Masucci, M. G. (2009b). Three Epstein-Barr virus latency proteins independently promote genomic instability by inducing DNA damage, inhibiting DNA repair and inactivating cell cycle checkpoints. *Oncogene, 28*, 3997–4008.

Guo, A., Salomoni, P., Luo, J., Shih, A., Zhong, S., Gu, W., et al. (2000). The function of PML in p53-dependent apoptosis. *Nature Cell Biology, 2*, 730–736.

Gurrieri, C., Capodieci, P., Bernardi, R., Scaglioni, P. P., Nafa, K., Rush, L. J., et al. (2004). Loss of the tumor suppressor PML in human cancers of multiple histologic origins. *Journal of the National Cancer Institute, 96*, 269–279.

Holowaty, M. N., Sheng, Y., Nguyen, T., Arrowsmith, C., & Frappier, L. (2003a). Protein interaction domains of the ubiquitin specific protease, USP7/HAUSP. *Journal of Biological Chemistry, 278*, 47753–47761.

Holowaty, M. N., Zeghouf, M., Wu, H., Tellam, J., Athanasopoulos, V., Greenblatt, J., et al. (2003b). Protein profiling with Epstein-Barr nuclear antigen 1 reveals an interaction with the herpesvirus-associated ubiquitin-specific protease HAUSP/USP7. *Journal of Biological Chemistry, 278*, 29987–29994.

Howley, P.M. (2006). Warts, cancer and ubiquitylation: lessons from the papillomaviruses. Trans Am Clin Climatol Assoc *117*, 113-126; discussion 126-117.

Hu, M., Gu, L., Li, M., Jeffrey, P. D., Gu, W., & Shi, Y. (2006). Structural basis of competitive recognition of p53 and MDM2 by HAUSP/USP7: Implications for the regulation of the p53-MDM2 pathway. *PLoS Biology, 4*, e27.

Kamranvar, S. A., & Masucci, M. G. (2011). The Epstein-Barr virus nuclear antigen-1 promotes telomere dysfunction via induction of oxidative stress. *Leukemia, 25*, 1017–1025.

Kaul, R., Murakami, M., Choudhuri, T., & Robertson, E. S. (2007). Epstein-Barr virus latent nuclear antigens can induce metastasis in a nude mouse model. *Journal of Virology, 81*, 10352–10361.

Kim, H. S., & Lee, M. S. (2007). STAT1 as a key modulator of cell death. *Cellular Signalling, 19*, 454–465.

Lassoued, S., Ben Ameur, R., Ayadi, W., Gargouri, B., Ben Mansour, R., & Attia, H. (2008). Epstein-Barr virus induces an oxidative stress during the early stages of infection in B lymphocytes, epithelial, and lymphoblastoid cell lines. *Molecular and Cellular Biochemistry, 313*, 179–186.

Lear, A. L., Rowe, M., Kurilla, M. G., Lee, S., Henderson, S., Kieff, E., et al. (1992). The Epstein-Barr virus (EBV) nuclear antigen 1 BamHI F promoter is activated on entry of EBV-transformed B cells into the lytic cycle. *Journal of Virology, 66*, 7461–7468.

Li, M., Chen, D., Shiloh, A., Luo, J., Nikolaev, A. Y., Qin, J., et al. (2002). Deubiquitination of p53 by HAUSP is an important pathway for p53 stabilization. *Nature, 416*, 648–653.

Li, M., Brooks, C. L., Kon, N., & Gu, W. (2004). A dynamic role of HAUSP in the p53-Mdm2 pathway. *Molecular Cell, 13*, 879–886.

Lu, J., Murakami, M., Verma, S. C., Cai, Q., Haldar, S., Kaul, R., et al. (2011). Epstein-Barr Virus nuclear antigen 1 (EBNA1) confers resistance to apoptosis in EBV-positive B-lymphoma cells through up-regulation of survivin. *Virology, 410*, 64–75.

Murakami, M., Lan, K., Subramanian, C., & Robertson, E. S. (2005). Epstein-Barr virus nuclear antigen 1 interacts with Nm23-H1 in lymphoblastoid cell lines and inhibits its ability to suppress cell migration. *Journal of Virology, 79*, 1559–1568.

Nicholson, B., & Suresh Kumar, K. G. (2011). The multifaceted roles of USP7: New therapeutic opportunities. *Cell Biochemistry and Biophysics, 60*, 61–68.

O'Nions, J., Turner, A., Craig, R., & Allday, M. J. (2006). Epstein-Barr virus selectively deregulates DNA damage responses in normal B cells but has no detectable effect on regulation of the tumor suppressor p53. *Journal of Virology, 80*, 12408–12413.

Onnis, A., Navari, M., Antonicelli, G., Morettini, F., Mannucci, S., De Falco, G., et al. (2012). Epstein-Barr nuclear antigen 1 induces expression of the cellular microRNA hsa-miR-127 and impairing B-cell differentiation in EBV-infected memory B cells. New insights into the pathogenesis of Burkitt lymphoma. *Blood Cancer Journal, 2*, e84.

Pearson, M., Carbone, R., Sebastiani, C., Cioce, M., Fagioli, M., Saito, S., et al. (2000). PML regulates p53 acetylation and premature senescence induced by oncogenic Ras. *Nature, 406*, 207–210.

Reichelt, M., Wang, L., Sommer, M., Perrino, J., Nour, A. M., Sen, N., et al. (2011). Entrapment of viral capsids in nuclear PML cages is an intrinsic antiviral host defense against varicella-zoster virus. *PLoS Pathogens, 7*, e1001266.

Salomoni, P., Ferguson, B. J., Wyllie, A. H., & Rich, T. (2008). New insights into the role of PML in tumour suppression. *Cell Research, 18*, 622–640.

Saridakis, V., Sheng, Y., Sarkari, F., Holowaty, M. N., Shire, K., Nguyen, T., et al. (2005). Structure of the p53 binding domain of HAUSP/USP7 bound to Epstein-Barr nuclear antigen 1 implications for EBV-mediated immortalization. *Molecular Cell, 18*, 25–36.

Sarkari, F., Wang, X., Nguyen, T., & Frappier, L. (2011). The herpesvirus associated ubiquitin specific protease, USP7, is a negative regulator of PML proteins and PML nuclear bodies. *PLoS One, 6*, e16598.

Scaglioni, P. P., Yung, T. M., Cai, L. F., Erdjument-Bromage, H., Kaufman, A. J., Singh, B., et al. (2006). A CK2-dependent mechanism for degradation of the PML tumor suppressor. *Cell, 126*, 269–283.

Scaglioni, P. P., Yung, T. M., Choi, S. C., Baldini, C., Konstantinidou, G., & Pandolfi, P. P. (2008). CK2 mediates phosphorylation and ubiquitin-mediated degradation of the PML tumor suppressor. *Molecular and Cellular Biochemistry, 316*, 149–154.

Schaefer, B. C., Strominger, J. L., & Speck, S. H. (1995). The Epstein-Barr virus BamHI F promoter is an early lytic promoter: Lack of correlation with EBNA 1 gene transcription in group 1 Burkitt's lymphoma cell lines. *Journal of Virology, 69*, 5039–5047.

Sheng, Y., Saridakis, V., Sarkari, F., Duan, S., Wu, T., Arrowsmith, C. H., et al. (2006). Molecular recognition of p53 and MDM2 by USP7/HAUSP. *Nature Structural and Molecular Biology, 13*, 285–291.

Sides, M. D., Block, G. J., Shan, B., Esteves, K. C., Lin, Z., Flemington, E. K., et al. (2011). Arsenic mediated disruption of promyelocytic leukemia protein nuclear bodies induces ganciclovir susceptibility in Epstein-Barr positive epithelial cells. *Virology, 416*, 86–97.

Sivachandran, N., Sarkari, F., & Frappier, L. (2008). Epstein-Barr nuclear antigen 1 contributes to nasopharyngeal carcinoma through disruption of PML nuclear bodies. *PLoS Pathogens, 4*, e1000170.

Sivachandran, N., Cao, J. Y., & Frappier, L. (2010). Epstein-Barr virus nuclear antigen 1 Hijacks the host kinase CK2 to disrupt PML nuclear bodies. *Journal of Virology, 84*, 11113–11123.

Sivachandran, N., Dawson, C. W., Young, L. S., Liu, F. F., Middeldorp, J., & Frappier, L. (2012a). Contributions of the Epstein-Barr virus EBNA1 protein to gastric carcinoma. *Journal of Virology, 86*, 60–68.

Sivachandran, N., Wang, X., & Frappier, L. (2012b). Functions of the Epstein-Barr virus EBNA1 protein in viral reactivation and lytic infection. *Journal of Virology, 86*, 6146–6158.

Takahashi, Y., Lallemand-Breitenbach, V., Zhu, J., & de The, H. (2004). PML nuclear bodies and apoptosis. *Oncogene, 23*, 2819–2824.

Valentine, R., Dawson, C. W., Hu, C., Shah, K. M., Owen, T. J., Date, K. L., et al. (2010). Epstein-Barr virus-encoded EBNA1 inhibits the canonical NF-kappaB pathway in carcinoma cells by inhibiting IKK phosphorylation. *Molecular Cancer, 9*, 1.

Wang, Z. G., Ruggero, D., Ronchetti, S., Zhong, S., Gaboli, M., Rivi, R., et al. (1998). PML is essential for multiple apoptotic pathways. *Nature Genetics, 20*, 266–272.

Wood, V. H., O'Neil, J. D., Wei, W., Stewart, S. E., Dawson, C. W., & Young, L. S. (2007). Epstein-Barr virus-encoded EBNA1 regulates cellular gene transcription and modulates the STAT1 and TGFbeta signaling pathways. *Oncogene, 26*, 4135–4147.

# Chapter 6
# Conclusion

There is no doubt that EBNA1 is a central player in EBV infection and EBV-associated tumours. The contributions that EBNA1 makes to tumourigenesis are likely multifaceted. In part, the role of EBNA1 would be indirect in maintaining the EBV episomes, enabling the continued expression of EBV-encoded proteins and RNA molecules which themselves directly contribute to cell transformation. However, mounting evidence also points to a direct role for EBNA1 in EBV-associated tumours, which could be due to contributions to initial cell transformation and/or due to the ability to maintain the proliferation or survival of the transformed cells. More studies will be necessary to determine the mechanisms by which the EBNA1-associated cellular effects are manifested.

L. Frappier, *EBNA1 and Epstein-Barr Virus Associated Tumours*,
SpringerBriefs in Cancer Research, DOI: 10.1007/978-1-4614-6886-8_6,
© The Author(s) 2013